SpringerBriefs in Applied Sciences and Technology

PoliMI SpringerBriefs

Springer, in cooperation with Politecnico di Milano, publishes the PoliMI SpringerBriefs, concise summaries of cutting-edge research and practical applications across a wide spectrum of fields. Featuring compact volumes of 50 to 125 (150 as a maximum) pages, the series covers a range of contents from professional to academic in the following research areas carried out at Politecnico:

- Aerospace Engineering
- Bioengineering
- Electrical Engineering
- Energy and Nuclear Science and Technology
- Environmental and Infrastructure Engineering
- Industrial Chemistry and Chemical Engineering
- Information Technology
- Management, Economics and Industrial Engineering
- Materials Engineering
- Mathematical Models and Methods in Engineering
- Mechanical Engineering
- Structural Seismic and Geotechnical Engineering
- Built Environment and Construction Engineering
- Physics
- Design and Technologies
- Urban Planning, Design, and Policy

http://www.polimi.it

Marcello Modica · Elena Solero

Brownfield Transformation in Fragile Territories

An Interreg-Based Action Research

POLITECNICO
MILANO 1863

Springer

Marcello Modica
Milan, Italy

Elena Solero
DAStU
Politecnico di Milano
Milan, Italy

ISSN 2191-530X ISSN 2191-5318 (electronic)
SpringerBriefs in Applied Sciences and Technology
ISSN 2282-2577 ISSN 2282-2585 (electronic)
PoliMI SpringerBriefs
ISBN 978-3-031-09082-0 ISBN 978-3-031-09083-7 (eBook)
https://doi.org/10.1007/978-3-031-09083-7

This Springer imprint is published by the registered company Springer Nature Switzerland AG
The registered company address is: Gewerbestrasse 11, 6330 Cham, Switzerland

Foreword: For a Contemporary Inhabitability of the European Alps

Fields of investigation and leading studies of important processes of industrial decommissioning have focused almost exclusively on cities and/or urbanised metropolitan contexts. They tend to reveal and prevail the urban dimension of regeneration programmes, projects and actions (Bondonio et al. 2005; Dixon et al. 2007; Hollander et al. 2010). Far from heavily urbanised and infrastructurally developed contexts, a host of different territories situated along the arc of the European Alps have been silently modelled by heavy industry (steelworks, mining, electrical and chemical plants). With an ecological and functional footprint much broader than their actual dimension, they modified different layered landscapes, almost always immersed in contexts of a rich history and beauty, inhabited by often closed communities, in many cases a result of their challenging topography and accessibility. Manufacturing activities and processes, together with a prosperous social era of factories-communities that generated and benefited from a robust social and charitable welfare (Galuzzi 2022), heavily conditioned local economies, societies and environments. Similar to urban systems, they were slowly exhausted over time, leaving behind decommissioned and abandoned landscapes, relics of a vast, broad and complex industrial modernity; the legacy of these decommissioning processes presents limits and problems for the adaptive reuse of contexts with varying conditions of fragility: economic (the closure of company-towns), social (the spaces of shrinkage) and environmental (polluting waste from manufacturing processes) (Galuzzi et al. 2020).

The European research *Alpine Industrial Landscapes Transformation (trAILs)*, funded within the Interreg *Alpine Space Programme* between 2018 and 2021, investigated this specific typology of decommissioning. Each country participating in the research (Italy, France, Austria and Slovenia) identified an area/district of a different original manufacturing typology (respectively in Borgo San Dalmazzo, L'Argentière-la-Bessée, Eisenerz, Tržič). They were then positioned within its multi-scale context of reference with the ambitious objective of reactivating economies, societies and ecologies, largely concealed and brusquely interrupted by decommissioning processes: a challenge for the regeneration of these territories, inspired by a mapping of decommissioned sites across the European Alps developed at the Technical University of Munich (Modica 2019), leader and coordinator of the trAILs research group. A decisive role in structuring the coordination of the trAILs research was played by Udo Weilacher, Landscape Architect, Author and Professor of Landscape Architecture.

This emerging territorial question, and simultaneously a latent potentiality and resource, as clearly explicated in this book, documents not only places and processes, in-depth and with broad evaluations. It argues and proposes solutions for promoting inclusive devices of governance focused on promoting a new inhabitability of territories, suggesting the fragile spaces of the Alps as true operative laboratories for testing the *Green Deal* promoted by the European Commission (European Commission 2019). This is, especially, in relation to four aspects:

- the capacity for local communities to be an integral part of transformative processes of regeneration (Alessandrini 2019);
- regeneration as a tool for contrasting climate change (Magni 2019);
- a temporary and incremental approach to regeneration, also as a means for experimenting with new economies tuned to local contexts;
- the definition of scenarios of regeneration that enable economic and social forces as an antidote to the impossibility to forecast the future (Wade 2012).

Enabling Local Communities

Community spaces are such only when communities interpret and model them. Processes of social innovation and bottom-up regeneration are fundamental resources for managing the complex changes that accompany transformations and a robust antidote to fragility. Fundamental for reducing the vulnerabilities of complex systems (Bertuglia and Vaio 2019), they are also an element for rehabilitating communities themselves (Vitale 2009; Blečić and Cecchini 2016; Vitillo 2022); not an abstract anti-fragility, but constructed from the specific historical, geographic, social and economic needs of territories and contexts, part of a process of social interaction implemented, together with the capital of a territory, to define appropriate local development. But how? Through the collective construction of pertinent biographies of resident communities, based on real needs that confront the concrete aspects of the life of communities and assist them in building programmes that reflect real situations and serve to orient actions. Consequently, supporting the enabling of local communities was an important and decisive part of the trAILs research, as described in different chapters of this book. This is due to the need to confront different legislation, diverse methods of organisation and inhomogeneous attitudes towards dialogue and participation, as the four experiences described in this book document so well.

Regeneration as a Project for Contrasting Climate Change

Any project for the future operates with the unavoidable awareness of the need to incorporate climate change and natural hazards within programmes (Beck 2000) and adopt an approach consistent with the multi-scalar nature of risks (Russo 2015). We live in a new condition, particularly evident in Alpine environments, where climate change and natural fragilities, combined with degrowth/demographic and economic contraction, tend to be more visible.

Climate change, already dramatic, has made natural hazards more invasive. In European Alpine contexts, this is particularly true of water-related risks (flooding). Water and its use were the matrix of production in Alpine territories and the driver of historic factories. How can we contrast these risks while simultaneously promoting a new economic phase? By working in two different directions: searching for a new form of fluvial landscapes, designed to contrast conditions of risk, utilising materials and technologies with a low ecological-environmental impact (Acreman 2004); exploiting new remote sensing technologies, biochemical analyses, hydrological calculation and modelling (new hydrological sciences, ecological management). Regeneration programmes can thus drive new development, with particular reference to policies and actions to contrast climate change, the creation of diffuse and capillary ecological networks and the production of eco-systemic services: forests, in particular, more than simply being of value as landscapes and recreational spaces (Burkhard et al. 2012), can play a determinant role in managing water resources (purification of water and air), in maintaining biodiversity (refuge for animals and vegetation) and mitigating–regulating climate, thanks also to their capacity to capture and store carbon. Green and blue infrastructures (Arcidiacono and Ronchi 2020) can be interpreted as territorial devices that favour an ecological transition: reducing the fragmentation of natural habitats; restoring the conditions required by natural processes; increasing the level of biological diversity and auto-regeneration; creating corridors linking up with external habitats; increasing the level of resilience of the urban ecosystem by increasing the loading capacity and performance of the built and unbuilt environment, forming new environmental and historical-cultural geographies. This awareness became the foundation of the trAILs research and the guiding star of its work.

Favouring a Transitory and Incremental Approach to Regeneration

In many Alpine contexts, rich with history, memory, environments and landscapes, additions would be out of place (Legrenzi 2022): it is sufficient to capitalise on and reveal what exists, focusing on and promoting territorial patterns and layers that demonstrate a capacity for resilience; landscapes that evolve from day to day, the result of bonds between the territory, community and historical-built capital that, together with natural and environmental capital, represent a capital to be cared for (Marson 2016). We must patiently select resilient veins, assist, reveal and rediscover natural situations and expose and promote hidden values and lines of force, while seeking to incorporate them within a new and diverse order that begins with the knowledge and expectations of local and contextual knowledge. The system of signs and objects represented by cultural heritage express long-lasting territorial values, to learn from and to question, also with regard to the relations they establish with contexts of settlement and open spaces.

Breaking through once impenetrable barriers, such as those of old factories, presents a twofold potentiality: opening spaces once inaccessible to local communities, containers of memory and an environment that has recaptured its original spaces, recovering processes and forms that produce identity but also experimenting with new urban economies, testing possible lasting uses. The spaces we design must necessarily be temporary, non-specialised, poly-functional, hybrid and reversible. We must pursue a transitory approach to regeneration, rediscovering qualities, beauty and adaptability to change (Jullien 2012), as well as its necessary process-based forms.

In extreme synthesis, favouring a transitory and incremental approach to regeneration means to define design tools that can accompany the consolidation of lasting uses; working with the concept of a temporary and progressive identity, changing, shifting and always open thanks to its multiple and perspectival nature; utilising the concept of durability, intended as a contrast to material and functional obsolescence, capable of triggering processes through the centrality of community welfare spaces, true drivers of incremental regeneration. Regenerating and caring for the territory and landscapes also challenges us to identify new trajectories for the growth of employment, with the birth and reinforcement of industries that interpret socioeconomic changes with intelligence but also for a new ecologically oriented production capable able to drive the development of small and medium-sized local enterprises, cooperative and social, innovative and sustainable (Secinaro and Biancone 2020), that look towards economies rooted in context.

One of the scenarios forecast by the trAILs research, though defined for the four different case studies, is represented by the activation of temporary uses (start-ups), which stimulate priority interventions capable of triggering processes for the partial reuse of areas through minimal actions to promote/constitute a pattern of environmental/landscape networks and a system of public spaces that open up towards places.

Constructing Scenarios for Enabling Socio-economic Regeneration

The research, beginning with an understanding of reference literature and virtuous examples of participatory design and planning processes, and its interpretation put into practice by this book, indicates a sensibility and a feasible approach—more than a rigid model to be followed a-critically—to the definition of scenarios of regeneration, for enabling local economic and social forces. This process is articulated in three steps, amply documented in this book: reconstructing the shared know-how of places (assessment); identifying approaches to listening in the form of participatory workshops (the test-design workshop procedure); and defining pertinent and sited guidelines (planning recommendations).

I will focus for a moment only on the latter. In my opinion, they represent the most fertile methodological legacy of the research. Design suggestions are expressed through the definition of essential principles to be respected and performance to be guaranteed in successive programming and planning tools. This is possible through the construction of development scenarios and verbal-visual guidelines. They must be the result of a collaborative model oriented towards the construction of an idea of the future required to coordinate actions, in the search for the difficult combination and delicate balance between freedom of action and business, and the unavoidable governance of phenomena. In extreme synthesis, enabling platforms that consent the free expression of societies and economies, interpreted through reference frameworks, intended as frames for possible actions (flexible and changing over time) that identify minimum patterns, able to recreate the continuity of geographic and historical-cultural environments and rooted in contexts; as well as lines of resistance defined by that nature and character of places (Eco 1997), real opportunities as well as systems open to possible events, beginning with keys for understanding, with contrasting and contradictory perspectives, though integral to a condition in which local communities recognise themselves and look with awareness towards their future, with care and passion for their local roots, and from a vantage point firmly rooted in our contemporary era. In relation to this aspect, the trAILs research, and above all its critical and fertile interpretation indicated by the authors in this book—and not only in reference to its replicability and transferability—offers interesting stimuli, spaces for discussion and further investigation yet to be filled.

Milan, Italy

Piergiorgio Vitillo
Politecnico di Milano, Responsible
for the Italian Research Unit of trAILs

References

Acreman M (2004) Water and ecology. UNESCO, Paris
Alessandrini G (2019) Sostenibilità e capability approach. Franco Angeli, Milano
Arcidiacono A, Ronchi S (2020) Ecosystem services and green infrastructure. Perspectives from spatial planning in Italy. Springer, Berlin
Beck U (2000) La società del rischio. Carrocci, Roma
Bertuglia CS, Vaio F (2019) Il fenomeno urbano e la complessità. Bollati Boringhieri, Torino
Blečić I, Cecchini A (2016) Verso una pianificazione antifragile. Come pensare al futuro senza prevederlo. Franco Angeli, Milano
Bondonio A, Callegari G, Franco C et al (2005) Il riuso delle aree industriali dismesse in Italia. Trenta casi studio. Alinea, Firenze

Burkhard B, Kroll F, Nedkov S et al (2012) Mapping ecosystem service supply, demand and budgets. Ecol Ind 21:17–29

Dixon T, Raco M, Catney P et al (2007) Sustainable Brownfield Regeneration: Liveable Places from Problem Spaces. Blackwell, Oxford

Eco U (1997) Kant e l'ornitorinco. Bompiani, Milano

European Commission (2019) Communication on The European Green Deal. COM(2019) 640 final. European Commission, Brussels

Galuzzi P (2022) Ivrea e Olivetti: il futuro di un discorso comunitario?. Urbanistica Informazioni (300):28–33

Galuzzi P, Solero E, Vitillo P (2020) Alpine space fragilities. A research line. Territorio (92):181–184s

Hollander J, Kirkwood N, Gold J (2010) Principles of brownfield regeneration. Cleanup, design, and reuse of derelict land. Islandpress, Washington DC

Jullien F (2012) Quella strana idea di bello. Il Mulino, Bologna

Legrenzi P (2022) Quando meno diventa più. La storia culturale e le buone pratiche della sottrazione. Raffaello Cortina Editore, Milano

Magni F (2019) Climate proof planning. L'adattamento in Italia tra sperimentazioni e innovazioni. Franco Angeli, Milano

Marson A (2016) La struttura del paesaggio. Laterza, Bari

Modica M (2019) Industrial Brownfield Sites in the Alps. A first Quantitative Overview and Potential Implications for Regional Development. J Alpine Res | Revue de géographie alpine 107(1)

Russo M (2015) Multiscalarità. Dimensioni e spazi della contemporaneità. Archivio di Studi regionali (113):5–22

Secinaro S, Biancone P (2020) La valutazione dell'impatto sociale. Aspetti metodologi. Pearson Education, Milano

Vitale T (2009) Introduzione: elogio del possibilismo. In: Vitale T (ed) Politiche Possibili. Carrocci, Roma, pp 14–20

Vitillo P (2022) Comunità insediate in azione. Urbanistica Informazioni (300):70–73

Wade W (2012) Scenario planning. A field guide to the future. Wiley, Hoboken, New Jersey

Preface

This book focuses on the emerging planning challenge of brownfield transformation and redevelopment in socioeconomic and environmentally fragile contexts and specifically in the European Alps. The main source of information and discussion, carefully reviewed, adapted and enriched, belongs to the EU Interreg project *trAILs—Alpine Industrial Landscapes Transformation*,[1] which run from April 2018 to May 2021 under the lead of Udo Weilacher and his team at the Chair of Landscape Architecture and Industrial Landscapes of the Technical University of Munich. The project, funded with 2.187.400 EUR within the 3rd call of Interreg Alpine Space Programme (2014–2020 funding period), set itself as a forerunning interdisciplinary and transnational initiative combining scientific research and local planning practice. This book, written and edited after the project end, aims to provide a concrete insight on the innovative planning aspects of the trAILs project, particularly on the content and outcomes of the Work Package T3 'Testing'. It focuses, in fact, on the specific 'test-design' method developed within the project to support local communities in defining the goals and steps of brownfield redevelopment, described and interpreted through the lens of spatial planning. The book is structured around nine chapters.

Chapter 1 introduces the topic of brownfield redevelopment in mountain regions and fragile territories, explaining in particular why the Alps are a valuable test-ground to study and tackle this challenge.

Chapter 2 describes the 'action research' approach behind the project operational logic, covering aspects such as the transnational and multi-expertise composition of the partnership, the roles of the involved institutions, the strategy of intervention and the selection of the pilot sites.

Chapter 3 briefly presents the test-design method as an innovative solution to address complex spatial transformations in fragile contexts, developed starting from established planning models such as test planning and participatory planning.

Chapters 4–7 focus respectively on the Austrian, Italian, French and Slovenian pilot sites, using the same structure to present and discuss the experiences, activities

[1] Project website: https://www.alpine-space.eu/projects/trails/en/home.

and preliminary results achieved by working on the sites and with the related local communities and stakeholders.

Chapter 8 brings forward a comparative review of the pilot-based experiences, evidencing the differences/similarities and the variations of the test-design approach used on-site. It discusses the adaption strategies according to the specific circumstances as well as the key elements shared by all the four experiences.

Chapter 9 finally points out some key conclusive remarks on the test-design methodology in the light of the project scope, discussing the characteristics, shortcomings and preconditions for a successful application and replication in similar contexts and situations.

Milan, Italy

Marcello Modica
Elena Solero

Acknowledgments

This book would have never seen the light without the approval of the trAILs project in the spring 2018 and the following three-year intense cooperation among scientists, experts and local communities from Austria, France, Germany, Italy and Slovenia. The participants belonged to ten institutions actively involved as project partners: Polytechnic University of Milan (POLIMI), Technical University of Munich (TUM), Vienna University of Technology (TUW), University of Verona (UNIVR), University of Ljubljana (UL), LAMORO Regional development agency (LAMORO), Architecture Urbanism and Environment of Vaucluse Department (CAUE 84), Registered association Styrian Iron Route (VESTE), Business support centre Kranj (BSCK), E-institute Ptuj (EZVD).

The making of this book in the post-project phase, a joint effort between the former project coordinator (at TUM) and the coordinator of the Work Package T3 'Testing' (led by POLIMI), is due to the scientific urgency to reflect on key methodological aspects associated with the test-design procedure developed in the project. In hoping that the content of this book will find the interest and appreciation of the whole trAILs team, besides the involved stakeholders, target groups and potential end-users, the authors would like to thank personally all those friends and colleagues from partner institutions who jointly realised this transnational project and made possible to reach its outstanding achievements: Piergiorgio Vitillo (POLIMI), Paolo Galuzzi (POLIMI), Gabriele Pasqui (POLIMI), Udo Weilacher (TUM), Johannes Kollmann (TUM), Katharina Strobl (TUM), Kerstin Bär (TUM), Markus Bauer (TUM), Lorenzo Migliorati (UNIVR), Liria Veronesi (UNIVR), Gianluca Lanfranchi (UNIVR), Veronica Polin (UNIVR), Vincenzo Prete (UNIVR), Andreas Voigt (TUW), Julia Forster (TUW), Julia Pechhacker (TUW), Stefan Bindreiter (TUW), Michael Rinnerthaler (TUW), Isabella Schuster (TUW), Tomaž Pipan (UL), Manca Krošelj (UL), Naja Marot (UL), Thomas Kleitz (CAUE 84), Jean-Charles Gros (CAUE 84), Annunzia Trischitta (CAUE 05), Alain Mars (CAUE 05), Gerfried Tiffner (VESTE), Kornelia Lemmer (VESTE), Heidi Pichler (VESTE), Umberto Fava (LAMORO), Sonia Abluton (LAMORO), Valentina Curato (LAMORO), Helena Cvenkel (BSCK), Uroš Brankovič (BSCK), Selma Terčon (BSCK), Zlatka Zastavnikovič (EZVD), Darko Ferčej (EZVD). A very special thanks also to the experts and consultants who supported the project administratively and in terms of communication: Dorothea Palenberg, Oleksii Tupikin, Aldo Cristadoro, Roberta Marchesi, Davide Fortini and his team, Nicolas Chesnel and his colleagues from the Joint Secretariat of the EU Interreg Alpine Space Programme.

Contents

Chapter 1
An Emerging Territorial Question

This chapter introduces the territorial challenge of industrial brownfield transformation in peripheral and fragile contexts, assuming mountain regions and specifically the Alps as the territory of reference. The brownfield issue is then briefly described through different thematic focuses (space, economy, society, environment) and temporal perspectives (origins, current impacts, future potential), including key facts and figures that make the Alps a significant case study area. The chapter ends by presenting the aims and structure of the book.

1.1 Mountain, Peripheral and Fragile Territories

The European space, historically and culturally heterogeneous, is increasingly being characterized by profound territorial disparities. Beyond the traditional core-periphery model (Wallerstein 1974), which was based on the distribution of (economic) power and patterns of accessibility, the actual condition of territorial peripherality derives from the interaction between specific socio-economic, cultural, geographic as well as environmental factors (Kühn 2015). This implies a significantly much complex and fragmented distribution of these 'fragile' territories across the continent, in fact not relegated to the margins as *no man's lands* but often strongly interwoven—spatially, socially, economically and environmentally—with the most developed regions and growth poles geographically adjoining. The concept of "inner periphery" (ESPON 2017) seems to match this condition, as it explains territorial peripherality as poor relational connectivity driven by both spatial and aspatial factors. As such, these lagging territorial enclaves are equally found in urban, rural and intermediate regions, with a prevalence of those affected by low economic potential (around 80%) in non-urban regions and those with limited access to services of general interest (around 50%) in metropolitan areas. An interesting fact, though rather undervalued in the ESPON policy brief, is the apparent overlapping between these

M. Modica and E. Solero, *Brownfield Transformation in Fragile Territories*,
PoliMI SpringerBriefs, https://doi.org/10.1007/978-3-031-09083-7_1

inner peripheries and non- or semi-urban territories characterized by highly environmental sensitivity, such as mountain and hilly regions. In this sense, territorial fragility should consider also and foremost the specific natural conditions—hydrogeology, morphology, biodiversity—of these areas, and especially the reciprocal influence between those and local development trends and opportunities.

A forerunning European policy case that tries to combine territorial (development) marginality with environmental inherent fragility is the Italian Strategy for Inner Areas (Barca et al. 2014). The identification of poorly developed areas with extensive "landscapes of abandonment", characterized by increased hydrogeological and ecological instability due to retreating or polarizing human influence, can be indeed transferred to many sparsely urbanized territories across the continent, strongly exposed yet unprepared to global socioeconomic and environmental changes. Mountain regions perfectly depict this 'integrated' view on territorial fragility. As inner peripheries, mountain areas are indeed subject the economic influence of surrounding urban and metropolitan flatland regions (Perlik 2019). At the same time, as outstanding yet precarious human-natural ecosystems, these territories and their habitability are increasingly challenged by climate-related environmental changes—impacting here faster and more significantly than elsewhere (Grabherr et al. 2010). In a way, morphological and environmental fragilities on one side and socioeconomic weaknesses on the other side are mutually influencing in mountain contexts: while the firsts tend to limit economic development and thus social change, the seconds are impeding to generate enough resources to manage and co-habit with these fragilities. This dual condition of socioeconomic marginalization and environmental vulnerability make of mountain regions an advanced laboratory to study, test and implement innovative strategies addressing sustainable territorial development in the framework of global changes.

1.2 Brownfields in Mountain Regions, a Latent Resource

Regional development policies and research often associate mountain regions to productive and living systems of rural character, based on traditional activities mostly belonging to the primary sector and shaped by extensive natural and semi-natural land uses. In the European context, however, this condition is valid only for very remote or inner mountain areas, while most of the low- and middle-mountain ranges are indeed characterized by highly heterogeneous productive landscapes with significantly high shares of manufacturing and services (Nordregio 2004). In particular, the European mountain ranges have a long industrial history linked to the intensive exploitation and processing of the natural resources here found. Running water was initially used for mechanical purposes in textile and paper industries and iron forges as well, and then to generate hydroelectric power serving heavy chemical and metallurgical industries. Mineral deposits were equally fostering ferrous and nonferrous metallurgical industries, but also cement production. The abundance of timber turned mountain areas into hotspots for pulp and paper industries, and charcoal-ironmaking too. In many noticeable cases, mountain resources and the attached industries were key drivers of the industrial and economic growth of regions and even entire countries (Bätzing 2015), although this aspect has been largely 'removed' from the collective modernistic image of mountains as rural *arcadia.*

In mountain regions as elsewhere in Europe, traditional resource- and labor-intensive industries are being challenged by the evolving global economic, productive and trade conditions. However, when it comes to deal with such structural changes, mountain industries face major problems than their urban, more sophisticated counterparts. Mountain industry is rather mono-structural, strongly attached to specific locational factors (e.g. resources, accessibility, etc.) and thus highly dependent from external (even infra-regional) conditions (Gebhardt 1990). As soon as mountain heavy industries were hit by the growing independence from raw materials and energy sources—following the globalization of trades and the liberalization of energy markets as well—, the resulting shifting pattern of regional and global production revealed all their inherent lack of resilience, thus leading to a rapid and irreversible decline (Raffestin and Crivelli 1988; Müller et al. 2006). Traditional industries, such as textile and ferrous metallurgy, and 'new' sectors as well, such as electrochemistry and electrometallurgy (among which aluminum industry in particular), were equally hardly affected.

The consequences of this regional industrial shrinkage were, and still are, particularly severe locally, often far-reaching into the already fragile social, economic and environmental conditions. Depopulation due to outmigration from inner valleys and loss of cultural identity (Migliorati and Veronesi 2020), persistent economic decline of previous mountain-industrial regions (Dalmasso 2007), uncontrolled contamination and ecological degradation (Wirth et al. 2012) are common issues for deindustrializing mountain communities and territories. Most of these situations can be related, directly and very concretely, to the presence of former productive sites that were gradually abandoned and decommissioned (Fig. 1.1), resulting in so-called brownfields: "derelict or underused sites that have been affected by the former [industrial] uses of the site and surrounding land; may have real or perceived contamination problems; are mainly [found] in developed urban areas; and require intervention to bring them back to beneficial use" (Oliver et al. 2005).

These sites, and especially their long-term regeneration and redevelopment, pose an enormous challenge to the affected communities. Both spatial and aspatial conditions concur to shape the problematic handling of brownfields in mountain areas (Modica and Weilacher 2019). Due to the specific characteristics of the mountain environment—namely the relevant topographic constraints, the prevalence of extensive semi-natural open spaces and the scattered, undefined urbanization pattern—the decommissioning of industrial sites often generates extensive and multi-faced brownfields, whose spatial footprint often spreads far beyond the core productive facilities including the complex system of supporting infrastructures and related functional spaces. The brownfield is here less identifiable with a single 'dysfunctional' or 'vacant' area, as in traditional urban contexts, but it rather looks more as a 'thick' and polycentric structural element of wider territorial transect. Still too

Fig. 1.1 The former INDEL site in Domodossola, Italy. © Marcello Modica

little addressed in local and regional brownfield redevelopment policies and strategies (if existing), this challenging physical condition is indeed confronted with major economic weakness and limitations typical of peripheral territories. On the one hand, there is the imbalance (real or perceived) between the amount of investments required for the site transformation and the uncertainty in terms of completion, outcomes and returns. The high costs for the site preparation—including essential remediation and reclamation stages—and its further development—including conservation and reuse of existing built structures or building anew—are often burdening on small mountain communities lacking adequate financial and contractual capacities. Unless strongly supported by larger institutional frameworks, both in terms of policies and finances, these communities are unable to produce, on their own, visions and long-term strategies for such complex operations, thus making the whole less attractive for private investors. On the other hand, the limited range and/or scale of activities that can be really implemented on these sites, given the contextual conditions (low demand, few potential users, scarce accessibility, etc.), further lows the chance of a successful long-term reconversion.

However, brownfield sites represent not only an extraordinary challenge for the affected mountain regions, but they are also and especially a concrete and unique resource to 'rethink' the mountain territory as a contemporary space of coexistence between man and nature. As a resource, brownfields are at all the effects a key component, yet latent and undervalued, of the territorial capital of mountain areas.

Defined as the sum of biophysical and human factors, of given environmental conditions and socio-cultural capabilities of a certain territory, the concept of territorial capital (Camagni and Capello 2013) perfectly applies to disused industrial sites and transitional spaces. These are indeed a material, physical and 'built' resource, an awaiting space capable—in the best way—of supporting a sustainable productive relaunch of deprived mountain-industrial communities.[1] This achievement can take place by converting brownfields in regional 'specialized' clusters of innovative micro-economies (Dax 2019), e.g. including small-scale business activities in green economy sectors linked to local nature-based production chains, or alternatively in cultural and touristic hubs based on industrial heritage valorization and fruition (Lorenzetti and Valsangiacomo 2016).

Mountain brownfields are also a powerful social resource, both as holders of local identity—though of difficult acceptance locally and extra-locally (Combal 2018)—and as levers of improved social cohesion. With reference to the latter, in particular, brownfield redevelopment can help to balance territorial disparities between growing and lagging mountain areas—the so-called "mountain gentrification" (Perlik 2011)—by attracting (or bringing back) certain social groups and interests.

Last but not least, and possibly most relevant, brownfield transformation can realistically support and enhance the environmental regeneration of ecologically fragile and often compromised mountain inhabited landscapes. Mountain ecosystems are highly dynamic and characterized by outstanding biodiversity and a rich ecological stratification, but at the same time they are also subject to an increasing pressure by local human activities (urbanization, transport, intensive agriculture, energy production, etc.). After a long-time contribution to altering the mountain ecosystem, former industrial sites represent, through their inherent condition, the occasion to restore and regenerate the hosting environment. This can be realized not only by removing soil and water contamination within the brownfield site (although this is the first necessary step), but also by pursuing a large-scale reclamation of the altered ecologies related to the site itself—which means, to consider the site as actually a key node of wider artificial-natural systems, thus a future-oriented provider of valuable ecosystem services (Grêt-Regamey et al. 2008; Mathey et al. 2015). Mountain brownfields are therefore an environmental resource in terms of decontamination, provision of ecosystem services and, indirectly, mitigation of potentially harmful impacts of natural hazards (e.g. flooding, landslides, etc.) and unsustainable urbanization models (e.g. soil de-sealing).

Considering the outstanding and foreseeable value of brownfields for the mountain environment, understood here in the broad sense, the challenge of their transformation, reuse and redevelopment is worthy to be seriously and urgently addressed.

[1] The reference is here to the post-crisis economic relaunch of the European manufacturing sector, addressed by specific EU policies (European Commission 2014), as well as to visionary strategies for empowering the productive role of mountain regions within the European economic-territorial system (Bonomi 2010).

1.3 The Alps as a Significant Case Study Area

Among the many European mountain ranges, the Alps stand out not only for their imposing morphology, cultural landscapes and touristic 'fame', but also and especially in terms of development and widespread industrialization. Located in the core of Europe, at the crossroad of powerful economic regions and dynamic metropolises, the Alps were in fact pushed through a modernization process far earlier than any other mountain region. This led to the emergence of different forms and cycles of industrial development, which today make of the Alps one of the few existing mountain regions worldwide characterized by mature industrialization (Bätzing et al. 2005). At least four industrial development phases can be identified (Gebhardt 1990; Modica 2019), since the continental spread of the First Industrial Revolution. In the first phase (1850–1880), the industrial transition of the existing pre-industrial activities—such as ironmaking, lime/cement production and textile manufacturing—was driven by the joint action of imported new technologies (coke, steam engine) and new transport means (railway). As result, the emerging industrial activities started to concentrate in large valley floors and transit corridors. The second phase (1880–1960) coincides with the industrial boom of the mountains, thanks to the discovery and widespread use of hydroelectric energy (Kopp 1969; Dalmasso 2007). New energy-intensive heavy industries gradually developed even in remote and inner valleys, especially in the emerging sectors of electrochemistry and electrometallurgy (aluminum, ferroalloys). In the third phase (1960–1980), new production paradigms such as flexible specialization united to targeted regional development policies fostered the development, almost everywhere in the Alps, of light industries in low added value sectors (mechanics, electric appliances, apparels, food). This process mostly occurred through forms of decentralized industrialization, such as branch-plants (Gebhardt 1990) and industrial districts (Fortis 1999).

In the fourth and current phase (1980-), the once historically relevant labor- and energy-intensive industries experience a significant decline, which is partially compensated by a growth of services and advanced tertiary activities only in major Alpine urban centers and transit hub regions. This structural transition, which largely affects the most industrialized Alpine regions and valleys, is clearly reflected in the shift of the secondary sector employment (regionally) from 50 to 36% between 1975 and 2000, and even further down to 18% by 2016.[2] Compared to the national averages of Alpine countries, these numbers prove that inner-Alpine deindustrialization is equally locally intensive though slightly delayed of at least one decade. A recent research (Modica 2022) has first identified the presence, across the entire Alpine space, of 302 sites[3] in those traditional industrial sectors mostly affected by decline—ferrous and nonferrous metallurgy, building material industry, textile

[2] These figures (Modica 2019) are derived from a merge and comparison of data at the regional and national level (sources: Alpine Convention, OECD, European Commission, Swiss Federal Statistics Office), using the mountainous area within the Alpine Convention perimeter as territory of reference.

[3] The identification criteria included the surface (equal or above 5 ha), the temporal framework of origin (ca. 1850–1960) and the industrial sector of reference.

industry, chemical industry and paper industry. Of these, 159 are actual brownfields, being already closed or downsized (Fig. 1.2). The sectoral distribution of brownfields sees a predominance of textile industry (26 brownfields on 35 sites in total) and nonferrous metallurgy (24 on 35), while the less affected sector appears to be paper industry (12 on 47). Building material industry (for most part cement industry), ferrous metallurgy and chemical industry shows an average of around 50% of brownfields on the total respective sites. The geographical distribution is particularly striking, as the majority of brownfields is found along the Alpine border (prealpine valleys and foothills, with some exceptional cases in inner areas) in correspondence of peri-Alpine urban agglomerations, with exceptional rates in the Italian Alpine regions (133 sites) followed by the Austrian (74) and French (43) ones.

Given this scenario, it is reasonable to expect that in many Alpine regions and valleys the already significant amount of disused or underused industrial sites will tend to increase in the near future. For many of the actual brownfields, in addition, no clear strategies and foreseeable futures are yet existing. While this is partially imputable to the still missing recognition, by policy makers and scientists, of brownfield regeneration as a crucial issue in the sustainable development of the Alpine region, it is also due to the absence of innovative and context-based approaches to tackle this specific challenge. In this regard, and considering the mentioned figures, the Alps represent at all the effects and purposes a unique and valuable 'laboratory' to

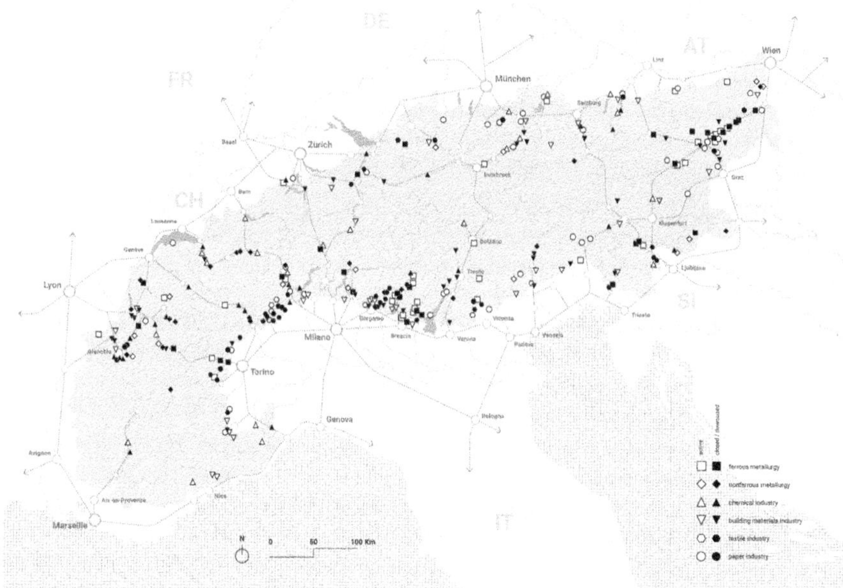

Fig. 1.2 Distribution of large industrial sites across the Alps as of 2021 (Modica 2022). Closed and downsized sites (actual brownfields) are black-filled shapes. © Marcello Modica

study brownfield redevelopment in mountain areas. This is particularly true if considering that these brownfields are somehow integrated into rather advanced socioeconomic systems—especially if compared to other European mountain regions—and a rapidly changing environment, highly sensitive to land use polarization processes. A novel approach to brownfield redevelopment in mountain contexts should start from these conditions, aiming to be inclusive (of space and communities), extensive (ensuring long-term sustainably) and transferable (accessible and replicable).

References

Barca F, Casavoli, Lucatelli (2014) A strategy for inner areas in Italy: definition, objectives, tools and governance. Materiali UVAL (31)

Bätzing W (2015) Die Alpen—Geschichte und Zukunft einer europäischen Kulturlandschaft. C. H. Beck, München

Bätzing W, Bartaletti F, Gubetti C (2005) Le Alpi. Una regione unica al centro dell'Europa. Bollati Boringhieri, Torino

Bonomi A (2010) La piattaforma produttiva nell'arco alpino. Economia trentina dossier (2/3):16–21

Camagni R, Capello R (2013) Regional competitiveness and territorial capital: a conceptual approach and empirical evidence from the European Union. Reg Stud 47(9):1383–1402

Combal C (2018) Bâti, patrimoine, industrie et identité : étude croisée de deux anciennes sociétés usinières Pechiney. Cahiers d'histoire de l'aluminium. J Hist Aluminium 1–2(60–61):103–121

Dalmasso A (2007) Industries et territoires dans les Alpes, XIXe–XXe siècles : tentative de typologie. In: Daumas J-C, Lamard P, Tissot L (eds) Les territoires de l'industrie en Europe (1750–2000). Presses universitaires de Franche-Comté, pp 87–102

Dax T (2019) Development of mountainous regions: smart specialization approaches as a means to overcoming peripheralization. In: Kristensen I, Dubois A, Teräs J (eds) Strategic approaches to regional development. smart experimentation in less-favoured regions. London and New York, pp 52–68

ESPON (2017) Inner peripheries: national territories facing challenges of access to basic services of general interest. ESPON

European Commission (2014) For a European industrial renaissance COM(2014)014. European Commission, Brussels

Fortis M (1999) Aree distrettuali prealpine. Meccanica, tessile, gomma e plastica. Franco Angeli, Milano

Gebhardt H (1990) Industrie im Alpenraum. Alpine Wirtschaftsentwicklung zwischen Aussenorientierung und endogenem Potential. Zugl überarb Habil -Schr, Univ Köln, 1987, Steiner, Stuttgart

Grabherr G, Gottfried M, Pauli H (2010) Climate change impacts in Alpine environments. Geogr Compass 4(8):1133–1153

Grêt-Regamey A, Walz A, Bebi P (2008) Valuing ecosystem services for sustainable landscape planning in Alpine regions. Mt Res Dev 28(2):156–165

Kopp H (1969) Industrialisierungsvorgänge in den Alpen. Mitteilungen Der Fränkischen Geographischen Gesellschaft 15(16):471–489

Kühn M (2015) Peripheralisation: theoretical concepts explaining socio-spatial inequalities. Eur Plan Stud 23(2):367–378

Lorenzetti L, Valsangiacomo N (2016) Alpi e patrimonio industriale. Alpes et patrimoine industriel. Alpen und industrielles Erbe. Mendrisio Academy Press, Mendrisio

Mathey J, Rößler S, Banse J et al. (2015) Brownfields as an element of green infrastructure for implementing ecosystem services into urban areas. J Urban Plan Dev 141(3)

Migliorati L, Veronesi L (2020) The consequences of modernity in the deep Europe: the transformation of industrial landscapes in Alpine regions. Italian Sociol Rev 10(1):1–29
Modica M (2022) Alpine industrial landscapes. Towards a new approach for brownfield redevelopment in mountain regions. Springer, Wiesbaden
Modica M, Weilacher U (2019) Alpine industrial landscapes in transition. Towards a transferable strategy for brownfield transformation in mountain regions. In: AESOP 2019—book of papers. IUAV, Venice, pp 147–165
Modica M (2019) Industrial brownfield sites in the Alps. A first quantitative overview and potential implications for regional development. J Alpine Res | Revue de géographie alpine 107(1)
Müller B, Finka M, Lintz G (2006) Rise and decline of industry in Central and Eastern Europe. A comparative study of cities and regions in eleven countries. Springer, Dordrecht
Nordregio (2004) Mountain areas in Europe: analysis of mountain areas in EU member states, acceding and other European Countries—Final Report. European Commission, Report
Oliver L, Ferber U, Grimski D et al (2005) The scale and nature of European brownfields. In: Oliver L, Millar K, Grimski D et al (eds) In: CABERNET. Proceedings of CABERNET 2005: the international conference on managing urban land. Land Quality Press, Notthingham, pp 274–281
Perlik M (2019) The spatial and economic transformation of mountain regions: landscapes as commodities. Routledge, London
Perlik M (2011) Alpine gentrification: The mountain village as a metropolitan neighbourhood. J Alpine Res | Revue de géographie alpine 99(1)
Raffestin C, Crivelli R (1988) L'industria alpina dal XVIII al XX secolo. Sfide e adattamenti. In: Martinengo E (ed) Le Alpi per l'Europa. Una Proposta Politica. Jaca Book, Milano, pp 161–184
Wallerstein I (1974) The modern world system. Academic Press, New York
Wirth P, Černič Mali B, Fischer W (2012) Post-mining regions in Central Europe. In: Problems, potentials, possibilities. Oekom, München

Chapter 2
Approach and Methodology

The chapter illustrates the background, rationale and approach of the Interreg trAILs project, focusing on four key aspects: principles of transnational cooperation and action research and their operational integration, composition and roles of the project partnership, structure and logic of intervention, selection and scope of the pilot sites. The innovative aspects of the project as well as its overall implementation are thoroughly described, thus providing a clear reference for understanding and interpreting the following methodological chapters.

2.1 Integrating Transnational Cooperation and Action Research

The transformation of industrial brownfield sites in fragile mountain contexts represents an unprecedented and complex challenge, certainly not suitable to be addressed through traditional urban redevelopment strategies. The special conditions of the context, before those of the site itself, must be taken into primary consideration when planning the transformation process as well as the expected outcomes. In this sense, the key question is here how to implement and develop a comprehensive regeneration strategy capable of re-connecting—physically and functionally—the brownfield site with its broader context, to ensure a sustainable future not only in economic terms but especially and foremost in social and environmental ones. Furthermore, considering the widespread distribution of these sites across the Alps and the expectable replication of many similar situations—regardless the local socio-cultural and legislative framework—, such a regeneration approach should also aim to be highly transferable, i.e. applicable through adaptation to a variety of cases and contextual conditions. The design and development of a specific 'planning' approach, valid and useful for all those mountain territories affected by the brownfield phenomenon, must therefore carefully balance the replicability issue (general) with the site- and context-related

M. Modica and E. Solero, *Brownfield Transformation in Fragile Territories*,
PoliMI SpringerBriefs, https://doi.org/10.1007/978-3-031-09083-7_2

challenges (specific). To be fulfilled, this objective requires the horizontal integration of external, transversal competences and local know-how, that is, of experts and communities. In this regard, European 'action research' funding schemes and especially transnational territorial cooperation programs offer a unique and outstanding opportunity for a concrete implementation.

With regards to the Alpine region, in particular, the EU Interreg Alpine Space Program (ASP)[1]—launched in 2000 and financed with ERDF funds through seven-year programming cycles, of which the last 2014–2020 having a budget of 139 million EUR—represents the main operational framework to implement and develop territorial cooperation (Interreg Alpine Space 2020). The aim of the ASP is, in fact, to foster the sustainable development of the greater Alpine region[2] through the thematic- and objective-based interaction between the diverse economic, social and environmental stakeholders from every institutional and societal levels (higher education and research, local and regional administration, business and private sector, public policies, NGOs, etc.). To guide territorial cooperation—implemented through consortia and partnerships focusing on specific development challenges—the program provides a series of strategic priorities, in line with the thematic objectives identified by the European Commission. In the previous 2014–20 programming, for example, the priorities were divided into: (1) Innovative Alpine Space (research and knowledge-transfer, territorial and social innovation), (2) Low Carbon Alpine Space (energy, transport and mobility, urban planning), (3) Livable Alpine Space (enhancement of natural and cultural resources, environmental protection) and (4) Well-governed Alpine Space (territorial governance, coordination with EUSALP[3]). The innovative aspect of the ASP, shared indeed by other similar European territorial cooperation programs, is precisely the ability to bring together and financially support geographically, culturally or thematically distant actors and institutions sharing similar intentions, interests and abilities regarding key territorial development challenges. Cooperation is meant, therefore, to generate new knowledge both individually and collectively: getting to know and experience what happens (or could happen) in neighbor regions, through an exchange of ideas, approaches and ways of doing, helps to generate—together—new strategies, transferable and replicable, to address transnational questions and issues. From a scientific point of view, the advantage of joining such a cooperation project is to have the chance to undertake *action research* (Stringer 2013), i.e. to proactively work on-site with those spaces, phenomena and actors primarily exposed to the challenge on stake and necessary

[1] Website of the Program: https://www.alpine-space.eu.

[2] The ASP covers an area of 450,000 km^2 which hosts around 70 million of inhabitants, inclusive of the orographic mountain range as well as part of the catchment areas of some of the major central European rivers (Danube, Adige, Po, Rhone and Rhine) and key metropolitan regions (Lyon, Geneve, Zurich, Munich, Turin, Milan, Vienna and Ljubljana).

[3] EUSALP is the European Strategy for the Alpine macro-region, a transnational platform joined by the major institutional stakeholders of the Alpine area (belonging to seven countries among which 5 EU member states, divided in 48 regions) aiming at shaping and implementing a common governance framework.

to define new methods and implement interventions. On the other hand, the territories, expressed through local and regional stakeholder networks, can profit from the participation to these cooperation projects by offering their own situation and experience as test-field for innovation, thus piloting change within their own extended community (e.g. similar territories, institutions or political frameworks).

The successful combination of transnational cooperation programs with action research, addressed by the trAILs project, is particularly striking and worth to be experienced. Action research is "a specific way of producing 'scientific knowledge' in which the production itself is inherently connected to the action on the field/territory and the communities inhabiting there" (Saija 2007: 49). Its scope is to determine changes on the reality by involving the belonging spaces and people, while reaching predetermined objectives by matching knowledge and action. In this sense, action research advances a radically different way of doing research compared to traditional models: it is not about doing research per se, but instead to be (part of) the research itself (Trombetta and Rosiello 2000). This approach fosters discussion and confrontation even among very different groups or beliefs, thus making of 'crisis' situations an opportunity for co-developing solutions—which is basically the scope of European transnational cooperation projects. Originally developed in the field of applied social research and then educational sciences, action research is increasingly raising the interest of technical-scientific disciplines, gaining meaningful and successful applications in urban research and urban planning studies. In this specific context, action research helps to overcome the old dichotomy between spatial/urban analysis and spatial/urban design. The territories as objects of investigation and study are here considered as 'communities of researchers', called to self-inquiry on the knowledge and interpretation of places, questioning given values and behaviors. Action research is thus at once collaborative and intervention-oriented research, an interpretative, qualitative way to fuel processes which does not limit itself to just valuing results. Theory does not constitute, in fact, the starting point (as in a top-down process) nor the arrival, but instead it finds its role as source of inspiration/definition for the praxis. The object of research is rooted in real situations and the participants themselves, who develop, conduct and shape the research through their 'daily' activities.

The action research process is founded on four main phases, which are repeating in loop until the expected results is achieved: planning, acting, observing, reflecting/evaluating (Cunningham 1976; Kemmis et al. 2013). These can be also grouped around three distinct moments—look, think, act—within the same cyclical process, aiming at reviewing and reinterpreting the procedures, including possible changes of direction (Stringer 2013). The 'planning' phase aims at verifying the needs, priorities and expectations, identifying the stakeholders to involve and formulating an action plan which sets time and resources. The 'acting' phase defines in detail the strategy to follow, the mid- and long-term objectives, the operational criteria and tools, as well as the single actions to undertake. The 'observing' phase includes both the monitoring of the process and the direct collection of information through notes, formulars, interviews, video-interviews, recordings, questionnaires, checklists. Finally, the 'evaluating' phase aims at confronting the preliminary and final outputs, reviewing and negotiating the project objectives while applying the necessary changes. Only after several intermediate evaluations, each corresponding at the end of a specific phase, the whole research could be completely assessed, perfectioned and thus disseminated. This four-phase logic has been considered while structuring and rolling out the trAILs project, though not as a straight procedure (which is not, indeed), but more to identify, classify and place specific actions and objectives within the research. The planning phase included the identification of pilot regions and sites, the selection of partners and observers, the definition of work packages as operational clusters of activities, each having a clear program and expected results. The acting phase included the further refining of work packages and the work plan based on the needs and capacities of the project team, as well as the step-by-step implementation and the active involvement of stakeholders and communities. The observing phase informed the listening process at the local and regional level, including fieldwork and site visits, meetings with local representatives, interviews and questionnaires, aiming at gaining both objective and subjective knowledge. The reflecting/evaluating phase included test-design workshops and post-workshop activities, review of outputs, feedback rounds with stakeholders, debriefing sessions, in some cases not foreseen in advance. By integrating these action research principles and operational character with transnational cooperation, the project could be equipped with the necessary structure and logic, hereafter explained in detail.

2.2 Interdisciplinary Expertise Meet Local Knowledge

In transnational research projects, the key characteristic of the partnership is certainly the different geographical and cultural backgrounds of involved participants. While the confrontation and exchange on shared territorial development issues and challenges is crucial to the definition of common strategies and approaches, in some specific cases it requires a further integration with different levels and types of knowledge. Considering the strong interdisciplinary and cross-scalar character of brownfield transformation—a complex process involving a broad range of expertise dealing with the built environment and its social, economic and ecological implications, as well as institutional levels and communities—the composition of the project partnership was designed accordingly (Fig. 2.1). The will was to integrate the scientific knowledge related to brownfield transformation with local and regional networks of stakeholders and institutions linked to some representative brownfield sites, acting as pilots for the project itself. The 'science cluster' of the partnership included research groups from different institutions and countries, each focusing on a key aspect of brownfield transformation in the context of mountain areas:

- the *Technical University of Munich* (TUM), involved through its *Chair of Landscape Architecture and Industrial Landscapes* (LAI), acting as project initiator and lead, and the *Chair of Restoration Ecology* (RE) covering the environmental aspects of the project. Founded in 2009 by Udo Weilacher, LAI deals

Fig. 2.1 The project partner institutions and the four pilot sites belong to different regions within the Interreg Alpine Space cooperation area. © Interreg project *trAILs*

with complex transformation processes in the contemporary landscape, focusing specifically on industrial and post-industrial spaces. Besides coordinating the project, LAI contributed to it by providing expertise and knowledge on landscape-based approaches to industrial brownfield transformation. Together with LAI, though independently working, RE contributed to the project through its specific knowledge on environmental regeneration and restoration of man-altered urban ecosystems, of which brownfield sites are particularly representative.

- the *Polytechnic University of Milan* (POLIMI), involved through its *Department of Architecture and Urban Studies* (DASTU). Established in 2013, DASTU carries out both theoretical and design-based research as well as educational activities on a strongly interdisciplinary base, considering its expertise in architectural and urban design, spatial planning and territorial governance, urban policies, preservation and intervention on built and natural heritage as well as architectural and urban history. With regards to the project, DASTU dealt with urban planning and design aspects of brownfield transformation, with a special focus on urban regeneration processes and strategic envisioning.
- the *Vienna University of Technology* (TUW), involved through its Department of Spatial Planning and especially its *Centre for Local Planning* (IFOER). Local planning focuses on the impacts that urban and territorial transformations have on planning policies, and therefore it strongly cooperates with local and regional authorities, irrespective of size, status and structure. Furthermore, IFOER has established an Interdisciplinary Centre for Spatial Simulation and Modelling

(Simlab) provided with the latest digital visualization tools and technologies for the built environment. Within the project, IFOER contributed with its proved expertise to the development GIS-based tools for Alpine brownfield mapping and assessment.

- the *University of Ljubljana* (UL), involved through its *Biotechnical Faculty* (BF) and specifically the Department of Landscape Architecture. Research at BF encompasses a broad set of topics including spatial and regional planning, territorial policy analysis, territorial cohesion and impact assessment, while counting a long-time experience in both national and transnational territorial cooperation. With regards to the project, in particular, UL-BF focused on brownfield-oriented policy analysis at the local and regional level, with a specific view on the aspects more influencing/influenced by urban and landscape transformation.
- the *University of Verona* (UNIVR), involved through its *Department of Human Sciences* (HS) in association with the *Department of Economic Sciences* (ES). The combination of HS and ES enables to deeply understand complex societal and economic systems, embracing expertise and providing knowledge in the fields of educational psychology, philosophy, cultural anthropology and sociology in the first, and economics and statistics in the second. Within the project, UNIVR contributed to identify and assess the social and economic impact of industrialization and de-industrialization on mountain communities, posing a particular attention on industrial cultural landscapes and industry-related territorial identity.

Provided with such a strongly interdisciplinary expertise, the 'science cluster' of the partnership was fully equipped to pursue an innovative and comprehensive study on Alpine brownfields as a unique spatial, environmental, social and economic challenge. Complementary to the research groups, four local and regional institutions were then involved in the partnership to bring actively into the project real territories and communities, with their own brownfield sites as test-ground. The selection of these regional partners was carried on in the early stage of the application phase through a 'top-down' approach, i.e. by identifying the suitable regions and sites first, as well as a "bottom-up" process based on networking and spontaneous candidacy. In the project, the regional partners acted as gatekeepers of their local brownfield sites (the pilot sites), addressing and supporting the mutual exchange between the regional institutions, stakeholders and communities and the interdisciplinary research cluster. The involved regional and local institutions were:

- the *Conseil d'Architecture, d'Urbanisme et de l'Environnement du Vaucluse* (CAUE 84), a public agency for architecture, urban and landscape planning created in 1979 and referring to the 132 municipalities of the Vaucluse department (Provence-Alpes-Côte d'Azur region, south-eastern France). As a public service association, CAUE 84 works daily with local authorities at various levels (municipalities, associations of municipalities, departments) providing consulting on built environment -related issues, as well as by supporting and integrating several projects of regional or national interest. With its great experience in brownfield redevelopment, CAEU 84 supported and coordinated in the project the twin agency CAUE 05, based in the Pays des Écrins department and thus directly responsible for the French pilot site in L'Argentière-la-Bessée (Hautes-Alpes).

- the *Agenzia di Sviluppo Langhe Monferrato Roero* (LAMORO), a regional development agency founded in 1995 by the initiative of 50 public institutions (40 municipalities and 10 chambers of commerce) based in the rural areas of southern Piedmont (north-western Italy). LAMORO activity is aimed at promoting a sustainable economy based on the capitalization of local territorial and human resources, including the enhancement of territorial identity and the promotion of innovative and inclusive models of governance. With its proved experience in EU territorial cooperation, LAMORO took over the overall responsibility and management of the Italian pilot site in Borgo San Dalmazzo and Valdieri (Valle Stura mountain district, Province of Cuneo).
- the *Verein Steirische Eisenstrasse* (VESTE), a regional association founded in 1986 for the promotion of the territorial cultural and natural heritage of the Leoben district, in central Austria. The association, structured as a consortium of 18 municipalities joined by the State of Styria, covers a variety of topics and projects, from urban revitalization to industrial heritage and economic development. During the last few decades, VESTE succeeded to address more than 20 million of EUR of investments, mostly through EU funding, on the regional development of the deindustrializing Leoben district. With this experience, VESTE joined the project as coordinator of the Austrian pilot site in the former mining town of Eisenerz (Obersteiermark).
- the *Business Development Center Kranj* (BSC KRANJ), a regional development agency established in 1995 as a consortium of 18 municipalities. The agency act as a link between regional public and private stakeholders (municipalities, consortiums, SMEs, etc.) and the Slovenian national authorities, being responsible for the preparation and implementation of the regional development program as well as for the implementation of the key strategic development priorities in accordance with the regional development council. Thanks to is expertise in preparing and managing EU projects covering different fields, from spatial planning to local development and heritage, BSC KRANJ took over the responsibility for the Slovenian pilot site in Tržič (Gorenjska / Upper Carniola non-administrative region).

With their profiles, knowledge and expertise, the four regional partners placed their activity and role at the interface of their own local and regional communities, the academic and research partners (the 'science cluster') and the EU program institution as well. In a way, the regional partners acted within the project consortium as true 'biographers' of their communities, contributing to the project goals and activities not only through a specific brownfield site as the object of transformation, but also involving and unfolding a local network of interests and capabilities.

Both the scientific partners and the regional partners formed, together, the operational core network of the project, carrying on day-by-day activities and implementing the path to the objectives and results. As extended network, many other institutions directly linked to the partners were also involved as observers. Their role consisted mainly in providing feedbacks to their related partners on the project activities, being also actively involved in specific activities, as well as to 'acquire' the project results and the generated knowledge for either scientific, institutional or interest purposes. The following observers were involved:

- *Ständiges Sekretariat der Alpenkonvention—Segretariato permanente della Convenzione delle Alpi—Secrétariat permanent de la Convention alpine—Stalni sekretariat Alpske konvencije* (international), associated to TUM;
- *Commission Internationale pour la Protection des Alpes—CIPRA International* (international), associated to TUM;
- *Bayerischen Staatsministerium des Innern, für Bau und Verkehr—Abteilung Wohnungswesen und Städtebauförderung, Sachgebiet Städtebauförderung* (Germany), associated to TUM;
- *Regione Lombardia—Direzione Generale Territorio, Urbanistica, Difesa del Suolo e Città Metropolitana—Unità Organizzativa Programmazione territoriale e urbanistica* (Italy), associated to POLIMI;
- *Regione Piemonte—Direzione Ambiente, Governo e Tutela del territorio—Settore Territorio e Paesaggio* (Italy), associated to LAMORO;
- *Région Auvergne Rhône-Alpes* (France), associated to CAUE 84;
- *Région Provence-Alpes-Côte d'Azur—Direction de l'aménagement du territoire et de la transition énergétique* (France), associated to CAUE 84;
- *Ministrstvo za okolje in prostor—Direktorat za prostor, graditev in stanovanja* (Slovenia), associated to UL;
- *Erz und Eisen Regionalentwicklungs Gmbh—LAG Steirische Eisenstrasse* (Austria), associated to VESTE;
- *Stadtgemeinde Leoben* (Austria), associated to VESTE;
- *Comunità Montana Valle Seriana* (Italy), associated to POLIMI;
- *Union Régionale des CAUE AURA* (France), associated to CAUE 84;
- *Universität Graz—Institut für Geographie und Raumforschung* (Austria), associated to TUW;
- *Università degli Studi di Bergamo* (Italy), associated to POLIMI;
- *ETH Zürich—Institute of Spatial and Landscape Development* (Switzerland), associated to TUM;
- *École Nationale Supérieure d'Architecture de Grenoble* (France), associated to CAUE 84;
- *Università della Svizzera Italiana—Accademia di Architettura di Mendrisio—Laboratorio di Storia delle Alpi* (Switzerland), associated to UNIVR;
- *TICCIH Austria—Österreichischer Denkmalrat für das kulturelle Erbe von Industrie und Technik* (Austria), associated to TUW.

As clearly visible from the list, the observers included mainly international 'framework' associations and institutions dealing with the Alpine region, regional authorities in the fields of spatial planning, environment and sustainable development, local authorities, universities and research centers specialized in aspects of brownfield redevelopment and industrial heritage. The strong interdisciplinary character of the project was thus also reflected in the observer network, confirming once more the relevance of cross-border, cross-level and multi-expertise cooperation.

2.3 Structure and Operational Logic

The composition of the partnership reflects in a way the operational logic behind the project. Differently from most transnational cooperation projects, where common challenges are tackled 'separately' through local piloting only (either public administrations, research centers or NGOs), the approach advanced by trAILs was to enable the whole project consortium to work proactively and collectively on all the pilot sites, using the same methodological and communicative framework. For this reason, the number of pilot sites was limited since the beginning to just four, a rather low amount compared to the average of Interreg projects.

Following this principle, the experts from the research institutes were first called to jointly develop a multi-thematic methodological framework to analyze and evaluate the current conditions of the site (assessment). Based on that, and with the active involvement of the local communities—coached by the regional partners –, the interdisciplinary expert group was in charge to outline the transformation potential of each site and to draw hand-on future scenarios (test-design). The interesting and innovative aspect of this two-step procedure was the horizontal, 'applied' involvement of the thematic expertise (covering spatial, environmental, social and economic aspects of brownfield redevelopment) across the four very diverse pilot sites. This helped, indeed, to gradually develop a specific methodology designed for the challenge on stake, continuously tested and adjusted site after site. Thus, the outcome was not generated from a later, detached comparison of the activities done, but indeed it was shaped step-by-step and on-site, through the involvement of the directly interested parties—exactly what is expected from 'action research'. Furthermore, the participation of each single regional partner in the assessment and co-design activities of the other pilot sites allowed a direct and constructive exchange of local knowledge among different regional and socio-cultural contexts. This aspect, really strategic in transnational cooperation projects, was fully integrated in the trAILs logic of intervention.

Due to the 'spatial' character of the problem on stake—brownfield transformation—, as well as considering the desired format of 'planning-based' solutions, the project introduced design as both a methodological and communication tool. The prospected sites transformation was first illustrated to the stakeholders and local communities by means of test-designs, i.e. speculative new spatial and functional layouts incorporating the main issues resulting from the previous assessments. Based on these design proposals as inputs, the discussion and evaluation with the stakeholders and communities, developed through a workshop format, led to identification of shared key principles (spatially defined) for the sites' future transformation. These key principles, finally, were turned into a set of alternative yet interlinked scenarios, the planning recommendations, a design-based communication tool serving (to the

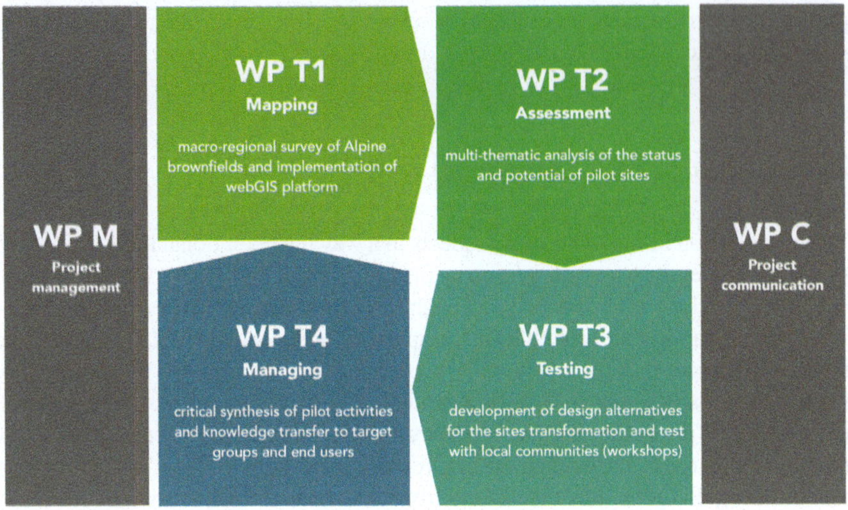

Fig. 2.2 The project implementation scheme. © Interreg project *trAILs*

communities) as a guide for the future management of the site. In this sense, the project intervention logic allowed planning and design languages to escape their own 'expert' perimeter and to turn into a meaningful and effective communication base for a wider public.

Operatively speaking, the project was structured around four work packages (WPs), as clusters of targeted activities aimed at specific preliminary results (Fig. 2.2). The first WP (T1 Mapping) focused on the implementation of a first Alpine-wide databased of brownfield sites, starting from the data available in the pilot regions and, as far as possible, extending to other regions within the cooperation territory. Under the coordination of TUW, the results of the mapping process, which involved different methodologies such as territorial surveys or codification of existing local or regional databases, were transferred and rendered visible into an interactive web-GIS platform. The second and third work packages constituted the operational core of the project, focusing specifically on the four pilot sites. The second WP (T2 Assessment) addressed the development of a transferable, comprehensive and multi-thematic assessment methodology for mountain brownfields, covering the different expertise involved (spatial and landscape planning, policy analysis, environmental sciences, social and economic sciences). The third WP (T3 Testing), closely connected to the previous one, focused on the development and implementation of a design-based methodology to evaluate the sites' future transformation with the active involvement of local and regional stakeholders. As core activity of this WP, a three-day planning workshop was set-up and carried on in all the pilot regions. The fourth and conclusive WP (T4 Managing) included the synthesis of project activities and the transfer of generated knowledge to end-users and target groups. Considering the project time frame, the four WPs were specifically designed to work 'in cascade', not just in sequence but slightly overlapping one over the next, to ensure a continuous flow of information from one activity to the other, as well as from one pilot site to the next. However, if the procedure is considered without the project time frame, it can also function as a continuous cycle: the Mapping provides the sites (identification and location) to the Assessment and Testing, whose outcoming methodology is further refined and adapted in Managing and thus transferred to other sites, to start the process again.

2.4 The Pilot Sites as 'Action Field'

In territorial cooperation projects, the exchange of know-how and practices among the partners and observers takes place through confrontation and mutual addressing of concrete challenges. As previously seen, within trAILs this basic approach was strongly reinforced through an innovative methodology seeking to integrate the different expertise of the partners with *and on* the involved territories (sites and regions). Motivated by the specific interdisciplinary and novel challenge of mountain brownfield transformation, this integrative approach made of site-based piloting activities both the conceptual and operational core of the research project. Conceptual as, by focusing on highly representative and recurring situations of Alpine brownfields, these activities allowed to identify the common characters of the challenge and thus how to target that specifically. Operational also, because as field-based activities, the work with the pilot sites constituted indeed the chance for the project consortium to test the proposed approaches and strategies actively and concretely. The pilot sites, each including one relevant brownfield site and the local/regional context of reference, were therefore assumed by the project as true experimental laboratories for co-designing transformation, and not as just case studies for mutual learning processes. The active involvement of the pilots in all the stages of the research, from the initial mapping towards the definition of planning recommendations and their comparative methodological synthesis, puts the territorial dimension (spatial, environmental, social and economic) at the center of the stage, as also required by the specific question on stake.

The selection of the pilot sites was done in accordance with the regional partners, though it was based on a preliminary existing mapping of Alpine brownfields developed at TUM in 2017–2018 (Modica 2019) as well as on a set of key criteria. Considering the limitation to only four pilots 'imposed' by the project proposed methodology, it was necessary to identify relevant working sites differing in terms of content and context, but at the same time highly representative of 'average' situations across the Alps. The first criterion was therefore the original content of the brownfield site, i.e. to include different typologies of traditional mountain industries with their own architectural, spatial and landscape footprints. The typologies were identified on the basis of their functional and historical relationship with the mountain environment: a cement factory (building material industry > mineral extraction), a blast furnace ironworks (ore-based steelmaking > mineral extraction), an aluminum smelter (electrochemical industry > hydropower) and a cotton spinning mill (textile industry > waterpower and hydropower). A second criteria was the current content (or status) of the site, which enabled to work with brownfields in different conditions or 'transition stages': from closing-down sites (pre-brownfield) to partially reused sites (post-brownfield), the four pilots represented the whole process. A third criteria was the local and regional context, aimed at studying and testing brownfield transformation within different Alpine cultural frameworks—including specific administrative, economic, social and environmental features. The pilot sites belonged therefore to four mountain regions in France, Italy, Austria and Slovenia, all differing at first but inherently sharing similar key characteristics of those contexts where mountain brownfields are found: mid-mountain areas with less attractive scenic landscapes ('ordinary' Alps), rather peripherical and with underdeveloped touristic sector. As such, the identified brownfield contexts expressed also very different forms of territorial fragility, emerging from the specific bio-cultural features of each region and locality. The integration of these criteria—site content, site conditions and regional/cultural context—allowed to identify and work with four highly representative sites, expressing not only the diverse territorial 'impacts' of brownfields in mountain areas, but also and especially the different ways in which these impacts are interpreted and managed (or attempted to). The identified sites were:

- Italcementi cement plant in Borgo San Dalmazzo (quarries and processing plant) and Valdieri (quarries). Status: closing down site, already partially inactive. Context: lower Gesso valley, Maritime Alps (Cuneo province, ITALY);
- Pechiney aluminum works in L'Argentière-la-Bessée. Status: partially reused site (small business) yet with large portions still derelict. Context: upper Durance valley, Dauphine-Cottian Alps (Hautes-Alpes department, FRANCE);
- OAMG blast furnace plant in Eisenerz-Münichtal. Status: long-time closed and largely reused site (small business) yet with many areas still derelict and ecologically compromised. Context: Eisenerz Basin, Styrian Northern Alps (Leoben district, AUSTRIA);
- BPT cotton mill in Tržič. Status: closed down site with already planned transformation. Context: upper Sava valley, Karawanks mountains (Gorenjska non-administrative region, SLOVENIA).

The sites, their regional contexts and the related challenges will be presented in detail in Part 2 (On site), in separate chapters.

References

Cunningham B (1976) Action research: towards a procedural model. Human Relat 29(3):215–238

Interreg Alpine Space (2020) Alpine space programme—cooperation programme 2014–2020

Kemmis S, McTaggart R, Nixon R (2013) The action research planner: doing critical participatory action research. Springer, Singapore

Modica M (2019) Industrial brownfield sites in the Alps. A first quantitative overview and potential implications for regional development. J Alpine Res | Revue de géographie alpine 107(1)

Saija L (2007) Prospettive di ricerca-azione nella disciplina urbanistica. Infolio. 19:49–52
Stringer ET (2013) Action research. Sage Publications, Thousand Oaks, California
Trombetta C, Rosiello L (2000) La ricerca azione: il modello di Kurt Lewin e le sue applicazioni. Centro Studi Erickson, Trento

Chapter 3
Test-Design, an Experimental Planning Method

This chapter introduces the test-design method developed and tested within the project, as a challenge-specific combination of key aspects of two established planning models: test planning and participatory planning. A brief introduction of the two is provided at the beginning, highlighting the respective pros and cons and the increasing use in brownfield transformation processes. The test-design method is then explained through its main phases, discussing the adaptability of the procedure, the involvement of stakeholders and decision-makers and the form and function of foreseen outputs. In conclusion, the positive linkups between the test-design method and the territorial fragility of mountain regions is also explained.

3.1 Informal Planning for Brownfield Transformation

One of the key objectives of the trAILs project was to test and define a specific planning methodology to cope with brownfield transformation in mountain regions. The active involvement of local communities, stakeholders and decision-makers was considered, since the very beginning, an unavoidable condition of success, especially due to the higher relevance of local interests and impacts of brownfield transformation in these contexts compared to traditional urban ones (Modica and Weilacher 2019). Accordingly, the project was set to pursue an informal planning strategy, one allowing collaboration among different levels and target groups as well as exploratory design activities towards inclusive and shared solutions. To do so, two informal planning models used in brownfield transformation processes were taken as reference, reviewed and incrementally combined and adjusted throughout the project: test planning and participatory planning.

M. Modica and E. Solero, *Brownfield Transformation in Fragile Territories*,
PoliMI SpringerBriefs, https://doi.org/10.1007/978-3-031-09083-7_3

Inspired by the experience of the Vienna's Donauinsel,[1] the test planning method has been developed at ETH Zurich by Prof. Bernd Scholl in response to the failure of formal planning instruments in tackling the increasingly complex spatial transformations of nowadays urbanized territories (Scholl 2017). The test planning method is an informal planning procedure specifically designed for spatial planning authorities and experts, which aims to produce a legally binding agreement for any spatial transformation through an interdisciplinary and multi-stage review of different project proposals (a simulated design competition). Through the years, and specifically in the Swiss context, the test planning method has been successfully applied to complex brownfield transformations such as the one in Attisholz-Solothurn (Scholl et al. 2013). Although ensuring a strong legitimation of the planning output—a concrete agreement among key decision-makers—, the test planning procedure difficulty adapts to peripheral and weak contexts, such as mountain areas, due to the lack of local-based planning experts and considering the significant amount of financial and time resources.

A second model assumed as reference by the trAILs project was participatory planning, which refers to a broad set of methods and practices developed in the 1960s and 1970s along the criticism towards rational planning. The participatory approach ensures the active involvement of citizens and communities in the planning process, fostering consensus-building on urban development and regeneration programs and projects as well as practical decision-making (Forester 1999). Differently from test planning, participatory planning models are not limited to planning experts and institutions, thus opening to the local society as a whole, while the aim is often set towards producing a shared vision for any urban or territorial transformation through a thematic-oriented listening process. In recent years, participatory planning has gained increasing interest and usage in brownfield transformation processes, both in the cases of public and private-owned sites (Tendero and Plottu 2019; Oliveria e Costa et al. 2018). The inclusiveness of participatory planning models is surely an advantage, especially in peripheral and mountain territories where community-legitimation is extremely relevant. At the same time, the usually non-binding character of the output (a vision) might limit its effectiveness in contexts where such complex transformations already struggle to take place.

By combining the two approaches, the trAILs planning experts identified a hybrid solution, adaptable to the different contextual conditions of the pilot sites and based on the following principles:

- the ultimate goal of the process is to agree on a set of key issues (minimum requirements) for the future development of the site, to be co-identified with local and regional stakeholders and planning experts in a workshop setting;

---

[1] In the 1970s the city of Vienna promoted a two-stage competition for the realization of a flood protection system along the Danube, which led to the creation of a new artificial island (Donauinsel) hosting today a multifunctional park system. The innovative procedure, led by Jakob Maurer, was based on the cooperation between external rival planning teams and administrative planning departments at the local and regional level.

- the identification and selection of local and regional stakeholders and planning experts is crucial, and should include/cover the basic planning expertise related to brownfield transformation (e.g. transport/infrastructure, contamination/environment, urban development, cultural heritage, etc.);
- the key issues emerging from the workshop are highlighted/visualized on a diagrammatic map accompanied by a short report, as reference for all the participants and related institutions;
- the planning recommendations derived from the workshop and based on the identified key issues should resemble more a planning agreement (few points but relevant for everybody) than a general vision for the site's future.

In this way, the project could benefit from both models while developing its own, challenge-specific approach. The latter is briefly explained in the following paragraphs.

3.2 Test-Design as a Process

Starting from the objectives of regenerating economically-weakened mountain contexts due to deindustrialization, of considering climate change impacts and risks on fragile environments, and of actively involving the affected communities by removing formal and functional planning barriers, the project embraced an innovative way of design-driven participation defined as 'test-design' (Galuzzi et al. 2020; Weilacher and Modica 2021). As an inclusive co-design procedure, test-design was set to be applied on the four pilot sites in Eisenerz (Austria), Borgo San Dalmazzo (Italy), l'Argentière-la-Bessée (France) and Tržič (Slovenia), though adapting to the local environmental, economic and social circumstances. Furthermore, with the aim to define a methodology replicable beyond those four sites, i.e. operationally transferable to a multitude of other sites across the Alps, the test-design procedure was organized in advance around a few key macro-phases: (1) trace grounded biographies of local communities, (2) set up a participatory process of explorative design, (3) outline possible and desirable scenarios, and (4) define design-based suggestions and recommendations for the site transformation. These phases, summarized in the Table 3.1 and explained in the paragraphs below, are temporally aligned though not binding in their respective content and approach—thus ensuring a high adaptability and flexibility in the implementation.

Trace grounded biographies of local communities

For each site, the procedure started with a detailed site visit and a collective draft of an interpretative assessment report of the site and the regional context of reference, characterized by multi-scalar and multi-disciplinary structure and content: socio-demographic, economic, spatial, landscape and environmental aspects, policy framework and legislation, urban and landscape planning strategies and restrictions. The report was prepared with the essential support of local development agencies, also

Table 3.1 The trAILs test-design methodology

Phases	Actors	Outputs	
		Documents	Description
(1) Multi-thematic assessment	Experts	Site assessment dossier	Collection, organization and illustration of the essential information on the site and context (spatial and landscape features, historic-cultural features, environmental and ecological conditions, economic and socio-demographic conditions, local and regional policies, planning restrictions)
		Biography of community (reports and videos)	Interviews and video-interviews to local stakeholders and communities, and definition of related profiles. Selection of stakeholders to be involved in the test-design workshop
(2) Test-desig workshop	Experts	Explorative designs and comparative matrix	Different explorative design proposals are prepared by partner universities through teaching studios, workshops or summer schools. The proposals are then examined through a comparative matrix to highlight peculiarities and features
	Experts + local communities	Poster of common agreed issues	A selection of design proposals, together with the site dossier (1), is reviewed by the stakeholders and experts with the support of facilitators, aiming at co-identifying common planning issues for the site transformation (poster-based)

(continued)

Table 3.1 (continued)

Phases	Actors	Outputs	
		Documents	Description
(3) Alternative development paths/scenarios	Experts	Alternative tranformation scenarios (draft)	The test design workshop activities and results are reviewed by experts and translated into draft transformation scenarios
	Experts + local communities	Alternative transformation scenarios (definitive) and key map	The scenarios are shared with the involved stakeholders for a feedback round, and then finalized. A key map summarizing the essential planning aspects for the site transformation (open space network, environmental and landscape system, historical and landmark buildings and structures to keep) is produced
(4) Planning recommendations	Experts	Planning recommendations	The key map, scenarios and test-design process are conveyed to the stakeholders and communities through a planning document providing shared principles, recommendations and future steps for the site transformation

involved as project partners, which have secured the active engagement of key stakeholders in the related territories. With the contribution of experts in social sciences (University of Verona), these actors were carefully selected and first approached through in-depth interviews, in order to better understand and focus on local needs, desires and expectations. In the end, this comprehensive assessment phase led to the definition of real and grounded biographies of the site-related communities, practically a dossier combining textual descriptions, images, diagrams as well as video materials (interviews) which portrays the people and places with which the test-design procedure will interact.

Set up a participatory process of explorative design

The second phase is organized around two activities of design-based inquiry, arranged in sequence: university-held design workshops and locally held test-design workshops. In the first one, the involved universities and research departments—Polytechnic University of Milan, Technical Universities of Munich and Vienna, Landscape Architecture School of Ljubljana—set up different formats of design-driven workshops with their own architecture and planning students, seeking to produce design explorations of how the project sites could be transformed and redeveloped. These designs were aimed not at prefiguring in detail the sites' future, though defining binding functions and forms, but on the contrary, they acted as 'design devices' to highlight those key relations and performances considered as essential to outline the transformative framework (environmental, infrastructural and built layers as minimal supporting system), while creatively reinterpreting the main features of the sites and the relations with the surrounding context. The design workshops generated several design proposals for each site, whose complex and different syntax was then explained and synthesized by means of a comparative matrix.

The outcomes of the design workshop, together with the interpretative assessment report, were then presented as inputs at the test-design workshops, where local communities and local and regional stakeholders were jointly involved in a participatory planning and design activity. Coordinated by professional facilitators, the test-design workshops became thus the occasion for experts and designers to meet and confront with local representatives, namely regional and local authorities, associations, citizens' representatives, technicians and planners. The scope of test-design workshop was twofold: on the one hand, submitting design proposals for the future site transformation to the local community to enable a joint and expert evaluation, that is, to stimulate discussion and confrontation on key aspects among different target groups and interests; on the other hand, to jointly identify and agree upon a set of transformation goals and development paths for the site, as pre-condition to their integration into existing local regeneration policies and strategies.

At the test-design workshop, the participatory planning approach was realized by setting up several working groups and roundtables, either thematic or interdisciplinary, in which mixing different expert profiles and local competences. The groups were called to review the design proposals, defining priorities of intervention and key transformation questions having as reference the site dossier (assessment report) and the comparative matrix of the proposals. The preliminary conclusions from each roundtable discussion were collected during the workshop itself on a strategic diagram—a draft site map synthesizing and geolocalizing the key elements emerged from the discussion—and then presented and discussed among a closing plenary involving all the participants. The goal here was to compare the different outcomes of the groups and summarize the identified elements: the result can be defined as a 'map of the community', a useful instrument to outline desirable and possible development scenarios, as well as design-oriented suggestions and recommendations later carefully prepared by the planning experts of the project team.

Outline possible and desirable development scenarios

The aim this phase was to identify potential development paths, that is, planning devices capable of enabling the following programmatic and design steps, which consciously incorporate uncertainty and risk. These development paths are indeed planning scenarios, which propose different regeneration drivers depending on the key issues co-identified with local communities during the previous test-design workshop. Regard-less of the site in the focus, the project addressed three main scenarios:

- activation through temporary uses (*start-up*), with priority interventions capable of triggering a partial reuse of the area through minimal actions of enhancement/upgrade of landscape and ecological networks as well as micro-systems of public spaces;
- intensive development (*back-to-city*), through the adaptive reuse of existing buildings and structures and/or the creation of a new urban fabric, foreseeing as urbanization matrix a system of public spaces and paths opening up to the surrounding context;
- extensive development (*back-to-nature*), through a gradual deconstruction of the site and the replacement of built-up spaces with semi-natural ones and post-urban environments;

These are performative and spatial configurations that are radically distant from each other, outlined starting from the relationships and performances identified in the key map—in which the key issues that emerged from the participatory workshop are represented and displayed, i.e. the main network of public spaces, the environmental and landscape system to enhance and reconnect to the context, historical buildings and heritage structures to be safeguarded. The three enabling scenarios are not strictly binding, of course, and can foresee intermediate solutions in the following implementation phases—though having as common ground the temporary activation. After being drafted, the scenarios and the attached design recommendations were presented to the stakeholders previously involved in the test-design workshop to get their feedbacks on reality-matching and feasibility.

Define design-based suggestions and recommendations

The very final result of the test-design procedure consisted in a set of design-based recommendations for the site transformation, expressed through the definition of principles to follow and performances to ensure during the subsequent implementation phases. In the hands of the local community and decision-makers, these planning recommendations helps to figure out the transformative potential of the site and so to select, in a later step, the preferred development path and how to reach/realize that. The recommendations generated from the test-design process act therefore as a flexible and open framework for action, allowing the expression of local societies and economies in defining, consciously, their own desirable future.

3.3 Scenarios to Cope with Fragility and Uncertainty

Likewise contemporary urban contexts, also fragile areas with low economic intensity have experienced in recent decades a real estate based territorial development, often in form of a chaotic accumulation of low-quality and short-term interventions. Starting from this consideration, the trAILs project proposed to use the test-design methodology as a way to incorporate, in brownfield transformation, the history, culture and 'materiality' of local territories, thus orienting the emerging development strategies towards concrete possibilities of intervention (Vitale 2009) and antifragile planning solutions (Blečić and Cecchini 2016). In other words, test-design supports local communities in mountain territories in assuming control of their own transformations, clarifying challenges and opportunities while setting priorities and co-developing useful action-oriented programs. This is particularly relevant if considering that in mountain areas the effects of climate change and natural fragilities are often summing up to demographic and economic shrinkage, thus increasing the relevance and impact of each single spatial transformation. Furthermore, there is a diffused consciousness that the time of all-embracing and stable scenarios, based on detailed long-term designs, is over: planning through scenarios is, on the contrary, a collective and open interaction among different interest groups, the result of a collaborative processes leading to build that idea of future needed to coordinate our actions and interventions (Blečić 2012). Scenarios meant as alternative future stories are indeed a true challenge for the contemporary urban and territorial project. This perspective seems to be effective, as it embraces manifold diverging programs while providing to the local communities a context and a trajectory on which self-identifying: a useful scenario is not one that portrays the future precisely, but rather the one which allows a community to learn and be responsible (Wade 2012); and which at the same time favors a capability approach (Alessandrini 2019). The inevitable assumption of the culture of uncertainty interprets any future-oriented thought not as an incognita, but as an opportunity and a resource. The present conditions of uncertainty in defining time and space calls for new approaches and logics, to follow and implement by means of guiding programs and projectual devices. The ultimate goal is to create favorable contexts for the metabolization of change, and for the acceptance of any upcoming modification. It is therefore not a question of defining different strategies based on alternative scenarios, but rather of building different alternatives of future starting from a common, participated base.

References

Alessandrini G (2019) Sostenibilità e capability approach. Franco Angeli, Milano

Blečić I (2012) Costruzione degli Scenari per la Pianificazione. Franco Angeli, Milano

Blečić I, Cecchini A (2016) Verso una pianificazione antifragile. Come pensare al futuro senza prevederlo. Franco Angeli, Milano

Forester JF (1999) The deliberative practitioner. Encouraging participatory planning processes. The MIT Press, Cambridge, MA

Galuzzi P, Solero E, Vitillo P (2020) Alpine space fragilities A research line. Territorio. 92:181–184

Modica M, Weilacher U (2019) Alpine industrial landscapes in transition. Towards a transferable strategy for brownfield transformation in mountain regions. In: AESOP 2019—Book of Papers. IUAV, Venice, pp 147–165

Oliveria e Costa S, Kellecioglu I, Weber R (2018) Developing brownfields via public-private-people partnerships. Lessons learned from Baltic Urban Lab. Nordregio, Stockholm

Scholl B, Vinzens M, Staub B (2013) Test planning—A method with a future. Swiss Federation, Office for Spatial Development (ARE); Canton Solothurn, Office for Spatial Planning, Bern and Solothurn

Scholl B (2017) Building actor relationships and alliances for complex problem solving in spatial planning: The test planning method. disP Plan Rev 53(1):46–56

Tendero M, Plottu B (2019) A participatory decision support system for contaminated brownfield redevelopment: a case study from France. J Environ Plan Manage 62(10):1736–1760

Vitale T (2009) Introduzione: elogio del possibilismo. In: Vitale T (ed) Politiche Possibili. Carrocci, Roma, pp 14–20

Wade W (2012) Scenario planning. A field guide to the future. Wiley & Sons, Hoboken, New Jersey

Weilacher U, Modica M (2021) Interreg trAILs—Alpine industrial landscapes transformation. Project handbook. Technische Universität München, München

Chapter 4
Eisenerz, Austria

The chapter presents the Austrian pilot site through the activities and achieved results in the framework of the trAILs project. A brief overview of both the regional context and the brownfield site is first provided, describing the main features and transformation challenges. Secondly, the identified potentials and problems are listed, as emerged from to the spatial assessment and the site visit. The chapter describes then the test-design phase, focusing on the workshop setting and methodological aspects and procedures. Finally, the planning recommendations for the site are presented, as outlined during and after the workshop.

4.1 Regional and Site Overview[1]

The eastern part of Upper Styria is shaped by the middle-mountain range of the Ennstaler Alps to the north-west and the heavily industrialized and urbanized Mur-Mürz valley to the south-east. Due to its topography, the region presents a highly polarized pattern of accessibility, with the main center of Leoben (24,680 inhabitants) and the surrounding urban areas well-connected to Vienna, Graz, Salzburg and Klagenfurt via the A9 and S6 motorways and several railway lines, while the mountainous inland is barely connected with local and regional roads. The closest airports are Graz and Vienna, approximately 45–90 min by car from Leoben. As a territorial administrative unit, the region of Upper Styria (Hochsteiermark) covers 3250 km^2 divided among 49 municipalities, hosting a total of 16,077 inhabitants. Over 75% of the region is covered by forests—largely cultivated in the past for ironmaking and now for timber industries—, with the remaining 10% covered by grassland for agricultural use and 15% for settlements (the last figure being half of the 30% of the whole Styria). The economy of the region used to rely on mining and heavy industry for most part of the last two centuries. Despite the ongoing deindustrialization since

[1] Information retrieved and adapted from *trAILs* Deliverable D.T1.3.2 (Lemmer and Tiffner 2018).

M. Modica and E. Solero, *Brownfield Transformation in Fragile Territories*,
PoliMI SpringerBriefs, https://doi.org/10.1007/978-3-031-09083-7_4

the 1970s, iron and steelmaking still represent a relevant economic sector, especially between Leoben and Kapfenberg. The leading Austrian steel producer Voestalpine holds several production sites in the Mur-Mürz valley, employing around 10,000 people in the whole region. The primary sector is limited to forestry, while tourism is relatively underrepresented and underdeveloped compared to bordering mountain regions. The most recent regional development strategies aim to foster an alternative tourist development—based on nature experience, outdoor recreation but also industrial and mining heritage—as well as to attract new green businesses in connection to the mountain resources and environment.

The old mining center of Eisenerz is one of the symbols of the region's industrial past as well as a key example of industry-shaped mountain landscape. Located at the feet of the Erzberg—the "iron ore mountain", shaped by centuries-old mining activities—, the working town of Eisenerz perfectly embodies the sharp industrial decline which affected the region in the recent decades. Besides the many empty buildings belonging to the old workers' settlements and the run-down mine premises, the brownfield of the Münichtal blast furnace plant precisely tells this story. Founded in 1899 by the ÖAMG (Oesterreichisch-Alpine Montangesellschaft, translated as "Austrian-Alpine mining and industry company"), the Eisenerz-Münichtal plant was one of the first in Austria to be equipped with a modern coke-fired blast furnace, using as raw material the iron ore from the nearby Erzberg. A second blast furnace was

Fig. 4.1 Overview of the Münichtal site at the feet of the Kaiserschild massif. © Gerfried Tiffner

added in 1911, and the ensemble was then completed in 1920–23 with the construction of a large gas-fired power station fired with the hot gas from the furnaces. After the stop of activities imposed by the late years of the Second World War, the complex never resumed the production. Part of the site was reused as central workshops for the mine until the late 1970s, when the old blast furnaces were demolished and the site opened to host new activities. In 1984 the multinational company Pilkington established a new windshields production facility on the southern premises of the site, an activity which lasted for less than twenty years. In 2010, again, an aluminum recycling plant was set-up by Alumelt, followed in 2012 by a scrap metal facility of Primaras.

The whole site, covering an area of 150,000 m^2, appears today as a highly heterogeneous, largely underused industrial area (Fig. 4.1). Recently built industrial halls (from the 1980s onwards) coexists with partially reused old buildings and some abandoned structures, among which the former gas power station from the 1920s. On the northern edge of the site, beyond a small creek descending from the nearby mountain, a huge slag heap belonging to the original blast furnace plant rises for several meters above the ground, containing heavily contaminated soils and particles. The future development of the site is clearly challenged by the fragmented ownership (around five different properties) and the limited accessibility/visibility due to the very remote location. Despite offering plenty of space and infrastructures, the site is persistently underdeveloped due to the difficulties in finding investors for long-term and innovative development plans, considering also the difficult regional context.

4.2 Spatial Assessment[2]

The assessment of the spatial conditions of the site and surroundings began with a site visit, which took place on 24–25 January 2019. In the site visit, organized by VESTE, the project partners explored the Münichtal site and its premises accompanied by local stakeholders and experts. The actual visit on the site was conducted by foot and lasted only for a few hours, due to the difficult weather conditions. During the tour, the local representatives (mayor of Eisenerz and site owners) provided detailed explanations of the conditions and development challenges of the site, while answering to the group's specific questions. On the second day, the project consortium met in Leoben to discuss and plan the assessment methodology, built upon the different thematic focuses and methods from each scientific partner. The spatial assessment was implemented off-site based on the direct observations on site, as well as by integrating various local planning documents provided by the regional partner. A second site visit took place in May, before the test-design workshop, to allow the partners to double-check their assessments and take more time to explore the site with better weather conditions. The spatial assessment gave relevance to the location/accessibility pattern of the brownfield site and on its built heritage in terms of existing structures. The results are

[2] Cf. *trAILs* Deliverable D.T2.2.3 (Solero et al. 2019).

presented below grouped into three analytical frameworks: environment, settlements and mobility. Further analyses were also done (e.g. land property, contaminated areas, etc.) and used in the workshop.

4.2.1 Environment

The brownfield is mostly characterized by buildings and paved, impermeable open spaces, while the immediate surroundings are either built-up or extensively natural (forested slopes). The site does not include any area with significant environmental qualities, but instead it revealed to host several contaminated spots. On the contrary, the wider context presents a diversified and rich environmental pattern with high ecological value, reflecting the regional character—Styria as the "Green heart" of Austria. The situation is portrayed in a specific map (Fig. 4.2) and described through its potentials and problems in Table 4.1.

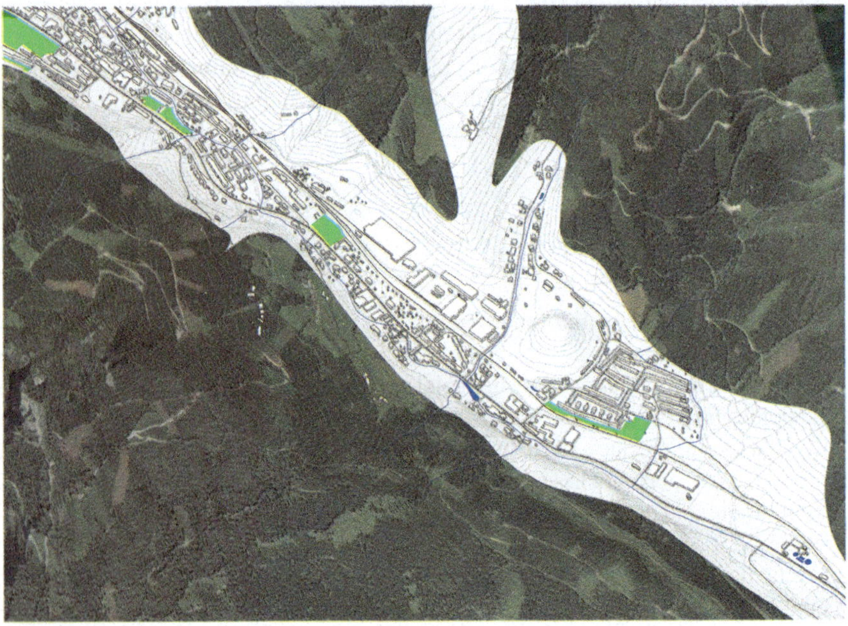

Fig. 4.2 Spatial assessment, environment map. © Politecnico di Milano/Interreg project *trAILs*

Table 4.1 Spatial assessment, environment-related potential and problems

Potentials	Problems
Landscape and environment, with high potential in terms of biodiversity and ecological connections as well as for touristic and recreational purposes (e.g. mining tour, Abenteuer)	Climate change and related risk of natural hazards such as flooding and avalanches (territorial fragility)
Abundance of wood as regional resource (approximately 90% of Eisenerz municipal territory is covered by forests)	Soil contamination on the northern edge of the site (former slag heap)
Natural protected areas in the vicinity (nature park/protected landscape)	
Große Fölz river separates the slag heap from the built-up area to the south	

4.2.2 Settlement Structure

The site is characterized by industrial buildings differing by age, conditions and use. A metal recycling company occupies around 55,000 m^2 in the middle of the site, with both reused and new buildings, while underused and abandoned areas are both to the north (former gas power station and premises) and the south of the site (former Pilkingtion plant). The settlement structure in the surroundings is characterized by a linear and rather compact development along the main road B115, with scattered residential areas on the mountain slopes. Many buildings are empty and in bad conditions, following the depopulation over the last decades. Notable is the old Münichtal workers' colony next to the slag heap, recently renovated as holiday apartments (Eisenerz Alpine Resort). The situation is portrayed in a specific map (Fig. 4.3) and described through its potentials and problems in Table 4.2.

4.2.3 Mobility

The accessibility of the site reflects that of Eisenerz, which is relatively poor. The site is located along the road B115, the historic "Eisenstraße" (iron road) which connects Eisenerz to Leoben (and thus Graz) to the south and Liezen (and thus Linz) to the north. The Leoben-Hieflau railway line that intercepts the site is used for freight transport only (very limited use), with no passenger service. Considering its position close to the town center, the site is well accessible via the local public transport network. The situation is portrayed in a specific map (Fig. 4.4) and described through its potentials and problems in Table 4.3.

Fig. 4.3 Spatial assessment, settlements map. © Politecnico di Milano/Interreg project *trAILs*

Table 4.2 Spatial assessment, settlements-related potential and problems

Potentials	Problems
The site is well embedded in the existing urban structure, nearby main services and cultural heritage spots (e.g. the medieval Kirchenburg St. Oswald)	Heterogeneous architectural quality of the site's building stock
Industrial heritage on the site (e.g. former power station)	Very high land ownership fragmentation
Proximity to the Erzberg Alpine Resort (possible touristic development of the site)	

In the view of the workshop, the spatial assessment was then integrated by the other thematic assessments carried on by the scientific partners, according to their own specialized expertise. The results are here briefly mentioned with reference to the thematic focus.

Socio-economic analysis[3] *(UNIVR)*: the local community is very cohesive and homogeneous in its composition, which brings risk of further closure to the outside/externals. The economic system lacks opportunities for young generations, fostering outmigration. A participatory approach to the site's transformation is

[3] Cf. *trAILs* Deliverables D.T2.2.4 (Migliorati et al. 2019) and D.T2.2.5 (Polin et al. 2019).

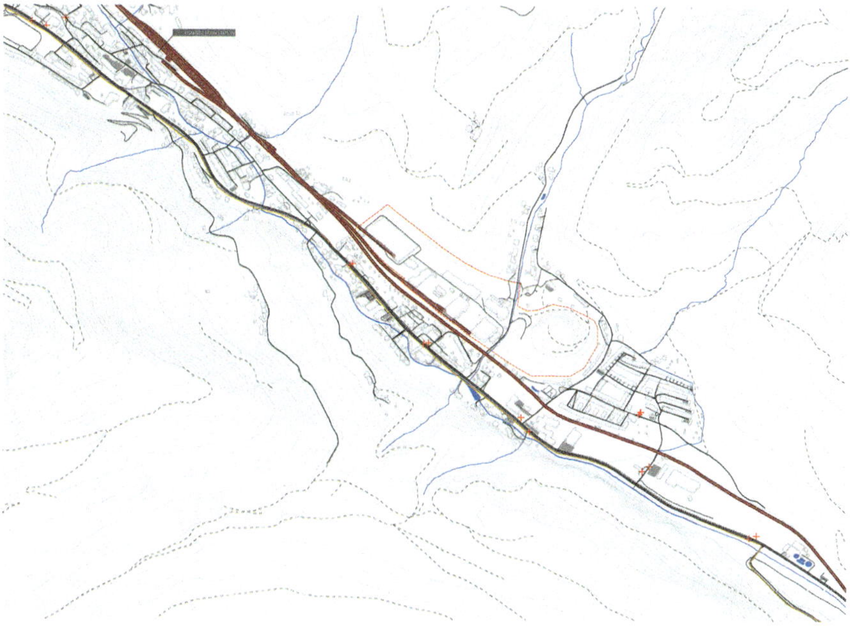

Fig. 4.4 Spatial assessment, mobility map. © Politecnico di Milano/Interreg project *trAILs*

Table 4.3 Spatial assessment, mobility-related potential and problems

Potentials	Problems
Proximity to the railway station (though not used), high potential for regional-wide connections by train (also for touristic purposes) even if there is no actual connection to Leoben (freight-only railway)	Lack of intra-regional mobility especially beyond daytime hours
Accessibility by road and bus lines, including seasonal public mobility for touristic purpose	Underdeveloped road network at the local and regional scale
	Underuse/obsolescence of the local railway line

strongly required to keep the community alive and ready to face the prospected changes.

Ecological analysis[4] *(TUM)*: the site features some interesting habitats for rare plants and some animal species. The reclamation of contaminated areas (tar deposits and slag heap) is highly recommended as well as the conservation and improvement of habitats including the restoration of the nearby river.

[4] Cf. *trAILs* Deliverable D.T2.2.6 (Strobl et al. 2019).

Policy analysis[5] *(UL)*: at the regional and local level, brownfield sites and especially their redevelopment are not perceived as an opportunity for sustainable urban and territorial regeneration, even if tools and incentives are existing but difficult to be applied.

4.3 Test-Design Workshop

Preliminary activities

The responsible project partner (TUW) set up the workshop agenda in advance, including a detailed meeting schedule (three days, of which the second one with stakeholders), inputs and expected outputs. The agenda was shared with the partners a couple of weeks before the event. On the first half day of the workshop, the project team performed a second, short site visit to the Münichtal site, to see those parts that were missed in the first site visit due to the bad weather conditions.

As design-based input material, useful to trigger the discussion on the roundtables, TUW provided four possible transformation scenarios or development pathways, drafted in the previous weeks together with their students in spatial planning. Furthermore, TUW proposed some working 'rules' to run the upcoming discussion with the stakeholders. In terms of organization, the agreed format was oriented towards having three thematic tables (i.e. working groups) focusing on specific topics (social and economic regeneration; environmental, ecological and landscape regeneration; spatial transformation and urban regeneration).

A detailed list of stakeholders to be involved was previously drafted, together with the regional partner VESTE. Around 50 among stakeholders and observers were contacted in advance, aiming at involving as more target groups as possible: site owners (different properties/areas), municipality (mayor, local council, head of administrative/planning sectors), representatives of other municipalities in the region having similar sites to deal with, town planners (especially the authors of the "Re-design Eisenerz" project), former workers on the site, representatives of major companies in Eisenerz (Erzberg mine) and regional civil society, such as the Leoben University, regional development agencies, chamber of commerce, tourist associations, culture associations, historians, heritage preservation authorities. A list of key questions for stakeholders was carefully prepared to stimulate the debate within the groups. Working materials to be arranged on the tables were also provided in advance.

Workshop settings

At the end of the first day, while discussing among the project team about the workshop organization of the following day (working groups with stakeholders), the planned setting based on thematic tables was changed in favor of interdisciplinary

[5] Cf. *trAILs* Deliverable D.T2.2.2 (Krošelj et al. 2019).

tables. This was motivated by the limited number of confirmed stakeholders (17[6]) as well as by their typology/sector. After the arrival and greeting of the stakeholders, the three interdisciplinary tables were arranged around three out of the four scenarios proposed by TUW. Each table was joined by 13 participants (both research partners and stakeholders, in a variable number) from different disciplines, expertise and sectors, including a moderator and a person appointed to live report the discussion.

Before the start of the table discussions, a short video produced by UNIVR[7] was played to be seen by all the participants. The video described the everyday life in Eisenerz, focusing on issues such local identity, the impact of deindustrialization, the current living conditions and the prospects or expectations for the future. The video helped the stakeholder to get engaged in the workshop activities. Furthermore, the three scenarios were briefly presented to the whole group too, together with the results of the assessment reports assembled by the project partners with the coordination of UL (policy analysis, territorial and spatial analysis, socioeconomic analysis, ecological assessment).

After these presentations, the working groups were invited to their respective tables and the discussion started. Each table was provided with the same working materials: a site dossier containing all the thematic assessments, a blank map of the site, a poster describing the proposed scenario (one per table), a booklet of the socio-economic assessment, a questionnaire for the environmental assessment. The list of key questions prepared in advance was used by the moderator to activate and guide the discussion. The results from the working tables were 'spatially' drafted on the blank maps using annotations, sketches and post-it, providing a first, rough visualization of the strategic ideas and emerged transformation paths (Fig. 4.5). These materials were reviewed on the third day by the project partners only, in a reflecting session aimed at reviewing the activities done and prepare the materials for the planning recommendations.

Roundtable 1

Group: 5 stakeholders, 8 research partners.

What do you associate with the topic of industrial brownfields? Have you already dealt with industrial brownfields in your professional environment?

---

[6] The participating stakeholders were: Sabina Cimerman (Government Of Styria—A17 Raumplanung), Werner Altrichter (site owner of Pilkington/Alumelt), Katharina Gugerell (Mining University Leoben), Werner Nussmüller (city developer, author of "redesign Eisenerz" project), Gunther Hasewend (former head of building dep. of Styria, initiator of "redesign Eisenerz"), Rainer Rosegger (sociologist; co-author of "redesign Eisenerz"), Christine Holzweber (Mayor of Eisenerz), Gerhard Niederhofer (local historian), Heidi Pichler (citizen + Eisenerz NGO on sport), Kornelia Lemmer (citizen + Eisenerz NGO on health), Dagmar Isele (culture association Innerberger Forum), Rudolf Haberl (citizen interested in regional development), three students (Eisenerz school of economy), Günter Speer (citizen of a neighboring municipality; expert in monument preservation), Wolfgang Fischer (University of Graz / observer).

[7] In cooperation with INTWIG communication and media agency.

- Brownfields are abandoned, unused former industrial sites characterized by contaminated soil and redundant built structures. An academic expert on brownfields and building recycling was present at the table.

Which one of the presented scenarios caught your attention the most? Which aspects did you like, which not? Which topics/issues have you missed?

- Although the preferred scenario was that of renaturalization, also considering the efforts of Eisenerz in becoming a touristic destination, the greenhouse proposal was perceived as an innovative and interesting approach.

What are your ideas for the transformation of the site?

- Adventure Park (in relation to the Erzberg Alpine Resort).
- Touristic or sports indoor activities (climbing walls, European sport centre for young people).
- Health uses (e.g. rehab clinics).
- Education, research and training activities.
- Temporary uses with cultural events.
- Production of eco-compatible building materials (e.g. wood or metal).
- Space for start-ups.

Which impacts arise with these new ideas for the site, the municipality and the region?

- Financial resources and their management (municipality, owner, investor), as well as funding programs (Styria, Austria, EU).
- Planning coordination and implementation.
- Diversification of economy (not only big companies but also small/medium companies).
- Enhancement of local resources (iron, wood, space, natural elements).

Which actors or organizations are needed for these new ideas? Which timeframe can we set?

- Site owners, municipality, investors, citizens, open mind-set people.

What was your expectation?

- To ensure new opportunities for young people and thus avoiding the shrinking city phenomenon.

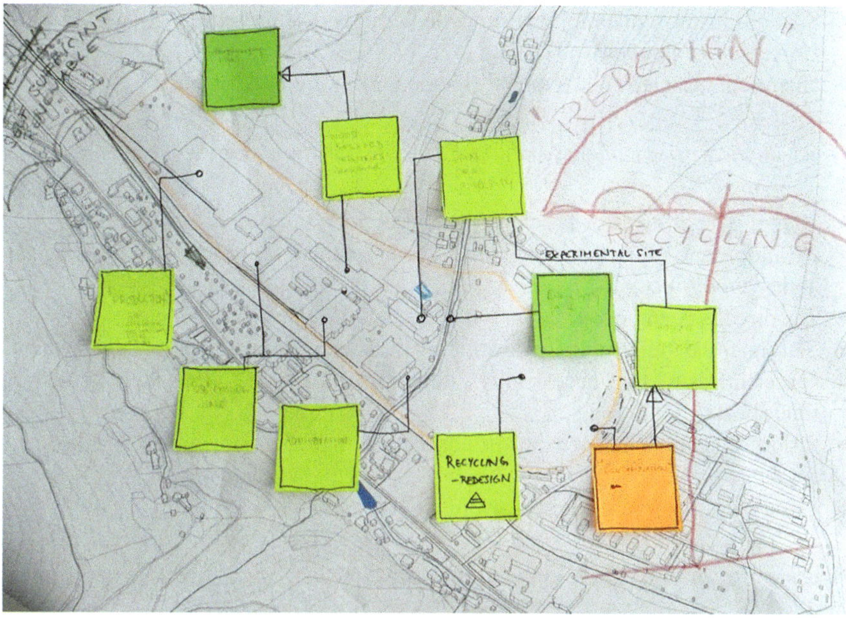

Fig. 4.5 Blank map with working annotations. © Julia Pechhacker

Roundtable 2

Group: 5 stakeholders, 8 research partners.

What do you associate with the topic of industrial brownfields? Have you already dealt with industrial brownfields in your professional environment?

- The topic of industrial brownfields involves inactive former industrial sites suitable for reuse/redesign/recycle. An academic expert on conversion of military, industrial and railway brownfields was present at the table.

Which one of the presented scenarios caught your attention the most? Which aspects did you like, which not? Which topics/issues have you missed?

- Re-design and recycling scenario, which consists on the one hand in the partial deconstruction/demolition of existing building (giving some areas back to nature), on the other in the establishment of new buildings characterized by a mixture of uses (small- and medium-sized enterprises).

What are your ideas for the transformation of the site?

- Research center (i.e. for material recycling).

- School cluster (education and knowledge in recycling sector) connected to the University of Leoben).
- Historical heritage enhancement (theme park 'mining heritage').

Which impacts arise with these new ideas for the site, the municipality and the region?

- Drawing up a 'investor-brochure' to attract potential businesses/investors.
- Development of a marketing and social inclusive strategy to raise acceptability/credibility.
- Tax directives and cooperation with regional development agencies.
- Setting up a detailed plan for developing the social and spatial potentials (available space, infrastructure accessibility, human resources including their know-how, time sheets).
- Strict connection between the mission statement of the municipality and the vision for the site.
- Good communication and involvement of locals.

Which actors or organizations are needed for these new ideas? Which timeframe can we set?

- Regional public administration, municipality of Eisenerz, University of Leoben, companies, media and communication tool.

What was your expectation?

- A sustainable development characterized by a high environmental and ecological approach.
- Rising awareness within the local community and stakeholders.

Roundtable 3

Group: 7 stakeholders (3 students), 8 research partners.

What do you associate with the topic of industrial brownfields? Have you already dealt with industrial brownfields in your professional environment?

- Industrial brownfields mean fallow lands resulting from industrial shutdown; vacancy problems; degraded buildings sometimes surrounded by high-quality natural space. At the table there were several experts in the topic of brownfields, wasteland and recovery policies.

Which one of the presented scenarios caught your attention the most? Which aspects did you like, which not? Which topics/issues have you missed?

- A new scenario derived by matching the renovation of the old buildings (e.g. old power substation) with a partially renaturation of the site.

What are your ideas for the transformation of the site?

- Sport/adventure activities.
- Cultural and touristic uses: creative sector, arts, exhibitions, galleries, industrial museum, touristic apartments.
- Start-ups and small enterprises.
- Research and education activities.
- Temporary uses (e.g. in the old substation): initially, spaces can be provided for free to activate area transformation.

Which impacts arise with these new ideas for the site, the municipality and the region?

- Stepwise transformation (to make it acceptable).
- Clear mission statement developed by the municipality: an official pathway is needed for an effective and shared transformation.
- Professional canvasser: a provider/operator on the site with high organization skills and able to coordinate investors, companies and inhabitants.
- Cost-benefits analysis.
- Great synergies between stakeholders, companies and citizens.
- Good communication and involvement of locals.

Which actors or organizations are needed for these new ideas? Which timeframe can we set?

- Eisenerz municipality financed by State and Federal government, owners, investors, stakeholders.

What was your expectation?

- A fruitful brainstorming, exchange of ideas, new practical and realistic solutions for the conversion of the site, as well as raising people's awareness.

4.4 Planning Recommendations

Eisenerz was the first test-design workshop to be held on-site at the regional level, serving therefore as experiment and test-ground for the other pilots. No key map was produced during the workshop, as the final discussion did not allow to merge the three scenarios into a shared single one. The key map, and thus the planning recommendations, were instead produced in the post-workshop phase, based on the critical review of the workshop outcomes by the project partners and experts. Besides the key map, which highlights the minimal shared requirements for the pilot area, three alternative 'development paths' were produced, ranging from the less invasive transformation (start-up/temporary activation) to the most radical ones ('back-to-city' and 'back-to-nature'). The 'back-to-nature' development path envisages a sort of sub-scenario named 'Greenhouse', from the related proposal presented by TUW. For each development path, the key actions related to the structuring issues for the site transformation—ecological network and public open spaces (including potentially relating built-up ones)—are described with text and a schematic illustrative plan.

Start-up/temporary activation

In view of a comprehensive transformation of the former industrial area, the reactivation of some of its parts might help, as a priority intervention, to trigger the whole regeneration process. The 'start-up' path envisages specific actions aimed at preparing the area, such as the preservation and gradual reuse of the industrial heritage structures connected to the former gas power station and its premises, as well as measures to open the area to the surrounding landscape. This latter objective involves the identification and development of a minimal ecological and mobility/open spaces network, in order to connect the activated parts of the site and the outer areas (Table 4.4; Fig. 4.6).

Table 4.4 Start-up scenario, key issues

Ecological network	*Minimal planting*: set-up a minimal network of open green spaces (tree-lined spaces, lawn areas) connecting the reactivated parts of the site
	Große Fölz restoration: remove built-up barriers within the riverbed or along the riverbanks and set a limit for new constructions or intensive activities along the river itself (river corridor); prevent the use of the river for energy production, since the river flow is insufficient to provide a substantial contribution to green energy improvements in the region (plus, micro-hydropower proves to have a significant impact on small river biodiversity)
	Green connections: connect green areas and the river corridor with in-site and off-site built areas, as well as to the hillside landscape
	Blue connections: enhance and better connect the river with the site, especially nearby the industrial heritage buildings
Public open spaces	*Former substation and public pool*: preserve and reuse the industrial heritage structures connected to the site 'historical memory' (landmarks) and restore the surrounding open spaces for recreational activities (pool/sauna) and cultural events
	New entrance: renovate the old building at the entrance of the site (new gate of the old factory)
	Public open spaces and pathway network: connect the activated parts of the site and enhance the livability of the area

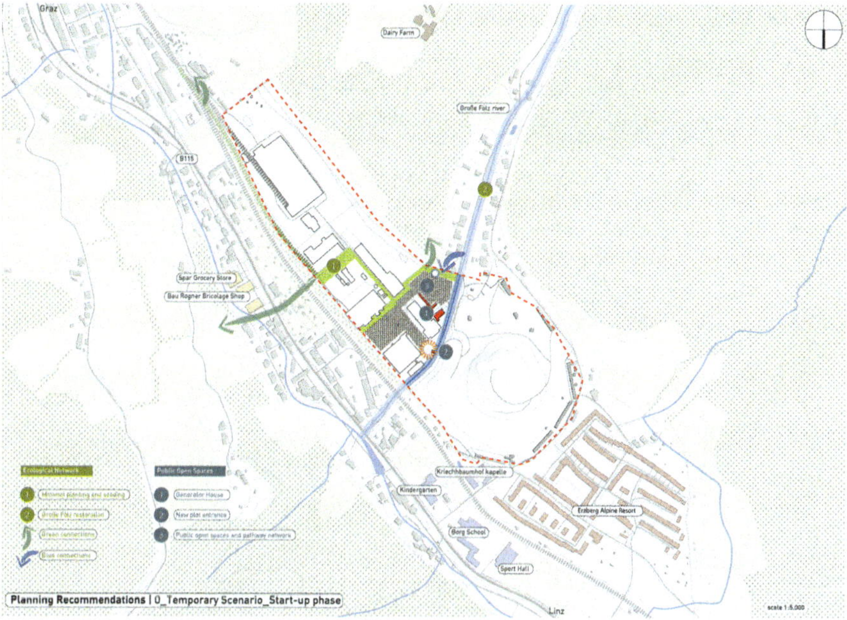

Fig. 4.6 Start-up scenario. © Politecnico di Milano/Interreg project *trAILs*

Intensive development/*back to city*

The 'back-to-city' development path addresses a substantial redevelopment of the site through the reuse and reactivation of existing buildings (either old or recent) as well as by defining new building areas and adding new volumes and built structures. To this purpose, the site internal layout is refurbished by setting-up of a network of new paths, public spaces and servitudes that integrates coherently the existing structure, while ameliorating the ecological conditions of the whole area (Table 4.5; Fig. 4.7).

Extensive development /*back to nature*

The 'back-to-nature' development path deals with the foreseeable perspective of a progressive underusage of the site. The conservation of just a few relevant built structures (identified as industrial heritage) coexists with an extensive environmental regeneration of the former industrial area, in coherence with the surrounding landscape. The renaturation of the site involves three main phases and relative regenerative landscapes: (1) sparsely vegetated landscape with spontaneous succession, (2) grassland-dominated landscape, (3) forest-dominated landscape. The sub-path 'back-to-nature_Greenhouse' addresses this ecological development with a less radical

Table 4.5 Intensive development scenario, key issues

Ecological network	*Minimal planting and/for remediation*: pay close attention to the tar contamination in order to avoid leakage to the groundwater. Favor succession processes on the slag heap and consequently led to increase flora and fauna
	Große Fölz restoration: remove built-up barriers within the riverbed or along the riverbanks and set a limit for new constructions or intensive activities along the river itself (river corridor); prevent the use of the river for energy production, since the river flow is insufficient to provide a substantial contribution to green energy improvements in the region (plus, micro-hydropower proves to have a significant impact on small river biodiversity)
	Enhancement of perimeter woodlands: improve the ecological conditions and the visual character of the woodlands along the western perimeter of the site, next to the mountain slopes
	Green railways tracks: renaturalize of unused railway tracks (as new ecological micro-corridors)
	Green roofs: improve the thermal and acoustic efficiency of the buildings and help to counteract the effects of waterproofing urban settlements
	Green connections: connect green areas and the river corridor with in-site and off-site built areas, as well as to the hillside landscape
	Blue connections: enhance and better connect the river with the site, especially nearby the industrial heritage buildings
Public open spaces	*Reuse of existing buildings and structures*: preserve and reuse the industrial heritage structures connected to the site 'historical memory' (landmarks) as well as the other, less relevant existing buildings
	Public open spaces and pathway network: connect the activated parts of the site and enhance the livability of the area
	New entrance: renovate the old building at the entrance of the site (new gate of the old factory)
	Tree-lined parking: provide for trees in and around parking lots to decrease pollutants and reduce temperature on the ground line. Planted areas also reduce storm water drainage problems, buffer wind and noise, and make passers-by and parking customers happy by providing cool shade
	New buildings: the redevelopment of the area implies the construction of new high performance and sustainable buildings to host new functions for the new settlement

approach, involving the reuse of some of the existing buildings and industrial halls as greenhouse (Table 4.6; Fig. 4.8).

Key Map

As shared between the three development paths, and preliminary to their future implementation, the key map integrates and illustrates the commonly agreed issues as identified in the test-design workshop. These include the essential network of public spaces, the environmental system and relevant ecological spaces as well as

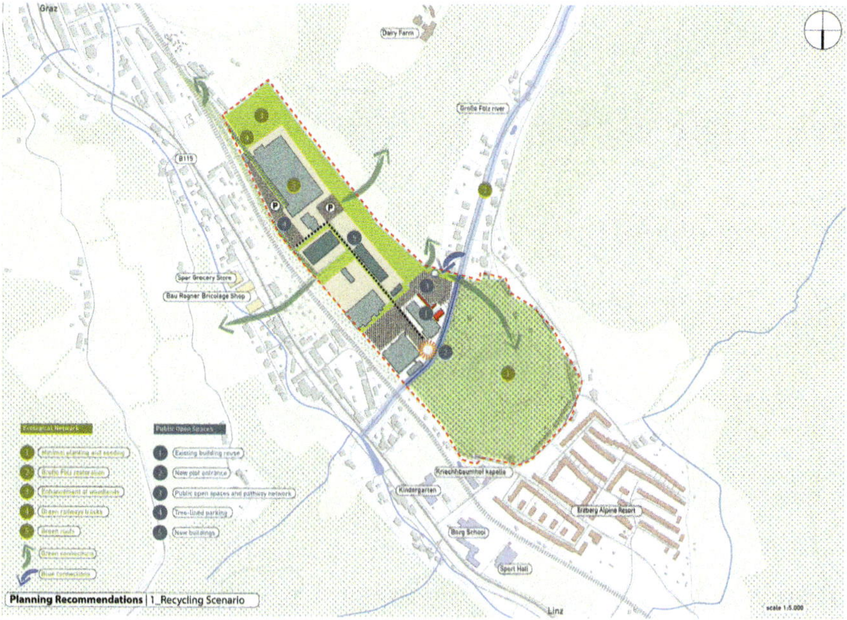

Fig. 4.7 Intensive development scenario. © Politecnico di Milano/Interreg project *trAILs*

Table 4.6 Extensive development scenario, key issues

Ecological network	*Minimal planting and/for remediation*: pay close attention to the tar contamination in order to avoid leakage to the groundwater. Favor succession processes on the slag heap and consequently led to increase flora and fauna
	Große Fölz restoration: remove built-up barriers within the riverbed or along the riverbanks and set a limit for new constructions or intensive activities along the river itself (river corridor); prevent the use of the river for energy production, since the river flow is insufficient to provide a substantial contribution to green energy improvements in the region (plus, micro-hydropower proves to have a significant impact on small river biodiversity)
	Enhancement of perimeter woodlands: improve the ecological conditions and the visual character of the woodlands along the western perimeter of the site, next to the mountain slopes
	Green connections: connect green areas and the river corridor with in-site and off-site built areas, as well as to the hillside landscape
	Blue connections: enhance and better connect the river with the site, especially nearby the industrial heritage buildings
Public open spaces	*Former substation*: preserve and reuse the industrial heritage structures connected to the site 'historical memory' (landmarks)
	New entrance: renovate the old building at the entrance of the site (new gate of the old factory)
	Public open spaces and pathway network: connect the activated parts of the site and enhance the livability of the area

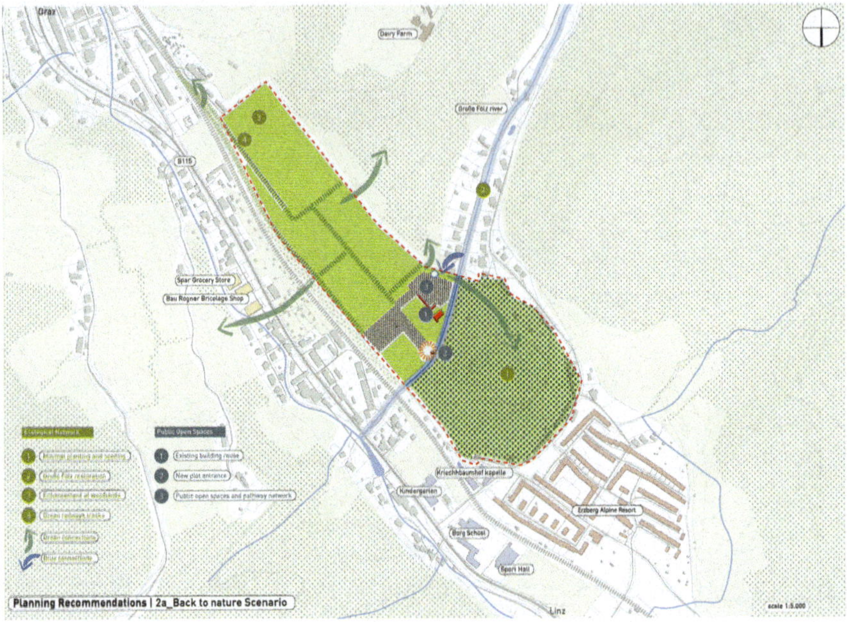

Fig. 4.8 Extensive development scenario. © Politecnico di Milano/Interreg project *trAILs*

the most relevant buildings and structures to be preserved and reused. By displaying the basic structure for the future development of the site, the key map is a valuable tool in the hands of the community, a starting point for any possible transformation. The minimum requirements for the site transformation and regeneration, no matter what will be the scenario that the community implement, are listed here below (Table 4.7; Fig. 4.9).

Table 4.7 Key map issues

Minimal planting and/for remediation: pay close attention to the tar contamination in order to avoid leakage to the groundwater. Favor succession processes on the slag heap and consequently led to increase flora and fauna
Große Fölz restoration: remove built-up barriers within the riverbed or along the riverbanks and set a limit for new constructions or intensive activities along the river itself (river corridor); prevent the use of the river for energy production, since the river flow is insufficient to provide a substantial contribution to green energy improvements in the region (plus, micro-hydropower proves to have a significant impact on small river biodiversity)
Enhancement of perimeter woodlands: improve the ecological conditions and the visual character of the woodlands along the western perimeter of the site, next to the mountain slopes
Green connections: connect green areas and the river corridor with in-site and off-site built areas, as well as to the hillside landscape
Blue connections: enhance and better connect the river with the site, especially nearby the industrial heritage buildings
Former substation: preserve and reuse the industrial heritage structures connected to the site 'historical memory' (landmarks)
New entrance: renovate the old building at the entrance of the site (new gate of the old factory)
Public open spaces and pathway network: connect the activated parts of the site and enhance the livability of the area
New building zones: define and delimit unbuilt lots for the site future development, including the necessary premises and servitudes

Fig. 4.9 Key map. © Politecnico di Milano/Interreg project *trAILs*

References

Krošelj M, Pipan T, Marot N et al (2019) EU Interreg project trAILs—Deliverable D.T2.2.2.—Policy assessment report Eisenerz. Available at: https://www.alpine-space.org/projects/trails/pilots/_results/wp-t2/trails_d_t2_2_2_policy_assessment-report_eisenerz.pdf

Lemmer K, Tiffner G (2018) EU Interreg project trAILs—Deliverable D.T1.3.2.—Pilot Profile Eisenerz. Available at: https://www.alpine-space.org/projects/trails/pilots/_results/wp-t1/pilot_profile_at.pdf

Migliorati L, Veronesi L, Lemmer K et al (2019) EU Interreg project trAILs—Deliverable D.T2.2.4.—Socio-demographic assessment report Eisenerz. Available at: https://www.alpine-space.org/projects/trails/pilots/_results/wp-t2/trails_d_t2_2_4_social_demographic_assessment-report_eisenerz.pdf

Polin V, Prete V, Lemmer K et al (2019) EU Interreg project trAILs—Deliverable D.T2.2.5.—Economic assessment report Eisenerz. Available at: https://www.alpine-space.org/projects/trails/pilots/_results/wp-t2/trails_d_t2_2_5_economic-assessment-report_eisenerz.pdf

Solero E, Vitillo P, Galuzzi P et al (2019) EU Interreg project trAILs—Deliverable D.T2.2.3.—Spatial planning and landscape assessment report Eisenerz. Available at: https://www.alpine-space.eu/projects/trails/pilots/_results/wp-t2/trails_d_t2_2_3_spatial_landscape_assessment-report_eisenerz.pdf

Strobl K, Bär K, Lemmer K et al (2019) EU Interreg project trAILs—Deliverable D.T2.2.6.—Environmental assessment report Eisenerz. Available at: https://www.alpine-space.org/projects/trails/pilots/_results/wp-t2/trails_d_t2_2_6_environmental_assessment-report_eisenerz.pdf

Chapter 5
Borgo San Dalmazzo, Italy

The chapter presents the Italian pilot site through the activities and achieved results in the framework of the trAILs project. A brief overview of both the regional context and the brownfield site is first provided, describing the main features and transformation challenges. Secondly, the identified potentials and problems are listed, as emerged from to the spatial assessment and the site visit. The chapter describes then the test-design phase, focusing on the workshop setting and methodological aspects and procedures. Finally, the planning recommendations for the site are presented, as outlined during and after the workshop.

5.1 Regional and Site Overview[1]

The southern half of the province of Cuneo, in the Italian north-western Piemonte region, is geographically and culturally shaped by the Maritime Alps and their anciently inhabited valleys. The pilot region covers two of the most important valleys, the Vermenagna and Gesso ones, both reaching the Alpine forelands in the municipality of Borgo San Dalmazzo (12,500 inhabitants). The other municipalities included in the region, and belonging to the two valleys, are Roccavione, Robilante, Roaschia, Valdieri, Vernante, Limone Piemonte and Entracque. Together, their total population is around 9500 inhabitants. In terms of accessibility, the international motorway E74 connects Borgo San Dalmazzo to Nice (France) following the Vermenagna valley and through the Colle di Tenda tunnel (1.3 km of length). From the French side, Borgo San Dalmazzo and its region can be also reached through the Stura Valley (SS21) via the Maddalena Pass (1996 mt) and the Lombarda Pass (2351 mt). The Cuneo-Ventimiglia railway line connects the Alps to the Mediterranean Sea via France. Besides Borgo San Dalmazzo, the territory is characterized by a pattern of small villages with a population density lower than the national level,

[1] Information retrieved and adapted from *trAILs* Deliverable D.T1.3.2 (Fava et al. 2018).

M. Modica and E. Solero, *Brownfield Transformation in Fragile Territories*, PoliMI SpringerBriefs, https://doi.org/10.1007/978-3-031-09083-7_5

an aged population and a low migratory dynamism. After the Second World War, the economy of the Gesso Valley underwent significant changes. Traditional activities such as agriculture and sheep farming lost their importance, while a process of impoverishment led to the gradual depopulation of the innermost valleys. At the same time, the presence of limestone and clay deposits fostered a rapid and intensive industrial development in the field of cement production and building material industries. Tourism adjoined the regional economic system in the 1980s, thanks to the outstanding environmental features of the local mountains. The Gesso valley lies in the middle of the Nature Park of the Maritime Alps, a 29,000 ha wide protected area which includes some of the highest peaks of the south-western Alps (e.g. the Monte Argentera, 3297 m). A second relevant protected area is the Natural Park of the Gesso and Stura rivers, which covers 5500 ha, 70 km of river systems while interesting a population of over 120.000 inhabitants distributed among 14 municipalities. Other touristic hotspots are Limone Piemonte in the Vermenagna valley (ski and winter tourism) and Valdieri in the Gesso valley (spa-town). The recent growth in the tourism sector is also profiting from the restructuring of local cement industries, namely thanks to the reduction of their heavy environmental and landscape impact.

One of the most significant industrial facilities in the region is the former Italcementi cement plant in Borgo San Dalmazzo, whose activity was linked to the construction and development of hydroelectric infrastructures and the economic recovery in the post-war period (Fig.5.1). The cement plant was established in 1941, though its troublesome completion during the war led to start operations only in 1947. The factory used the 'dry process' for clinker production and had an overall capacity of 150,000 t/y, upgraded to 450,000 t/y by 1956. The production complex included two large quarries ("Monte Cross" for limestone and "Terra Rossa" for clay shale), located along the Gesso valley between Borgo San Dalmazzo and Valdieri and connected to the cement plant by an aerial ropeway. For more than half a century, the Italcementi cement plant in Borgo San Dalmazzo was one of the most significant employers for the whole region. In December 2008, however, the economic crisis and the obsolescence of the plant led the company to decide the stop of clinker production. Strikes, protests and negotiations prevented the site closure. The plant was thus downsized and transformed into a grinding center, while the number of employees decreased from 97 in 2008 to 34 in 2015 and just 20 in 2019.

In the second half of 2019, the ownership of the plant (meanwhile inactive) was assumed by Buzzi Unicem, while the productive infrastructure (land and facilities) remained in the hand of Italcementi. Considering the future scenario that awaits the site, i.e. its physical and functional transformation, the involvement of the ownership in the definition of goals and priorities, as well as in the planning process itself, appears as a key challenge—especially due to the different expectations compared to the local community. Furthermore, the redesign of the former cement plant must consider the environmental restoration of the abandoned quarries located along the valley, thus pushing for a comprehensive and long-term territorial transformation strategy.

Fig. 5.1 Overview of the Italcementi cement plant from the Monserrato hill, towards the Vermenagna valley. © Marcello Modica

5.2 Spatial Assessment[2]

The assessment of the physical conditions of the site and its context began with a site visit, which took place on 27–28 June 2019. In the site visit, organized by LAMORO, the project partners were guided by local representatives and experts around the area of Borgo San Dalmazzo. The visit mostly focused on the premises of the Italcementi site, as the entrance permission to it was denied by the owning company. However, the group was allowed to see the site from the inside while remaining within the large apron beyond the entrance. On this occasion, a detailed description of the site was provided by TUM members, who had the chance to visit the site entirely the year before (as part of the design process). The site visit was conducted both by foot (site premises) and by bus (overview of the quarries in the upper Gesso valley), and lasted for more than half a day, being accompanied by detailed explanations by the local guides on the conditions and challenges of the site and its context. In the afternoon of the first day, the project team had the chance to deepen these issues with local stakeholders (mayors of Borgo San Dalmazzo and Valdieri, head of the urban planning department of Borgo San Dalmazzo, environmental expert

[2] Cf. *trAILs* Deliverable D.T2.2.3 (Solero et al. 2019).

from Legambiente Cuneo) in a roundtable session. The site visit and the following discussion were extremely useful to the partners to draft their thematic assessments.

Concerning specifically the spatial assessment, the knowledge gained on-site during the visit and the roundtable with the stakeholders was then integrated, off-site, with the local planning documents provided by the regional partners. Different aspects were covered, from the regional accessibility to the local conditions of the site and its context. The resulting spatial assessment is presented below through its three main analytical frameworks: environment, settlements and mobility. Further analyses were also done (e.g. land property, built heritage, etc.) and used in the workshop.

5.2.1 Environment

The site lies on the orographic left of the Gesso river, partially occupying the extended river bed in a situation of high ecological fragility and hydrogeological risk. The eastern portion of the site is completely unbuilt and includes a spontaneous woodland developed over a few decades. The western part is mostly built-up with production facilities and the surface is sealed (paved). The immediate surroundings are characterized by a mixed environmental pattern, with forested areas on the hill-sides and cultivated crops on the flatland. The larger context presents several nature

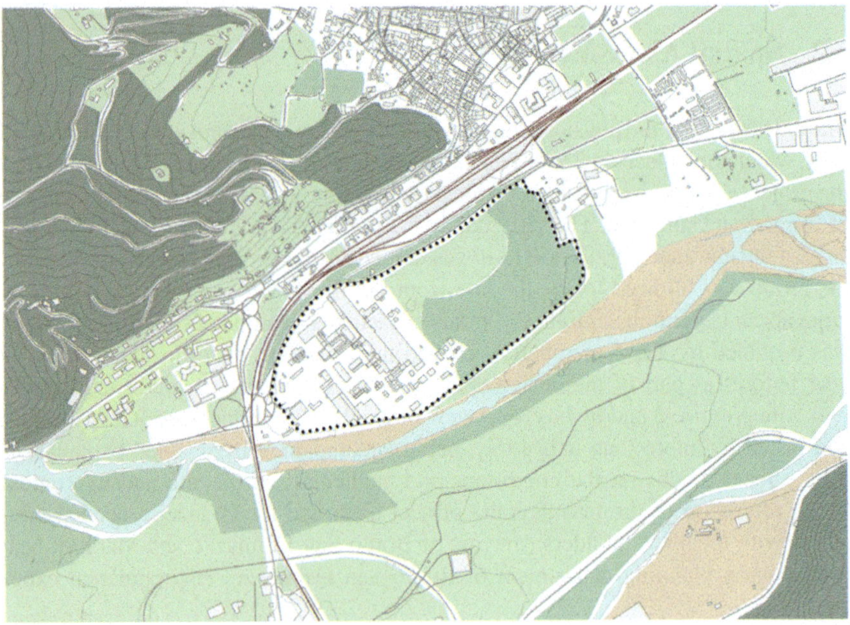

Fig. 5.2 Spatial assessment, environment map. © Politecnico di Milano/Interreg project *trAILs*

Table 5.1 Spatial assessment, environment-related potential and problems

Potentials	Problems
Diverse landscape, with high potential in terms of biodiversity and ecological connections as well as for touristic and recreational purposes	Climate change affecting the water regime of the Gesso, thus increased risk of natural hazards such as flooding for the site
Many forested areas representing a great potential for wood processing	Possibility of soil contamination at the site and in direct surroundings
Many natural or close-to-natural water bodies in the proximity of the site	Presence of large mineral extraction sites that negatively impact the landscape potential for tourism
Natural protected areas in the vicinity (national park and nature park)	
Location of the site along the Gesso, thus high potential for environmental valorization of the site in connection to the river corridor	

protected areas of high ecological value, a wide hydrographic network (mostly close-to-pristine) but also quarries and mining sites linked to the cement industry, some of which are posing environmental problems due to prolonged abandoned. The situation is portrayed in a specific map (Fig. 5.2) and described through its potentials and problems in Table 5.1.

5.2.2 Settlement Structure

Half of the site is filled up with massive industrial buildings (mostly unused) with rather extensive footprints. The industrial complex is separated from the surrounding urban pattern due to its location in the alluvial plane of the Gesso river, as well as because of the railway and regional road passing nearby. The settlement structure in the surroundings is characterized by scattered residential areas (except for Borgo San Dalmazzo old town) and small-scale industrial zones between the river and the railway. The situation is portrayed in a specific map (Fig. 5.3) and described through its potentials and problems in Table 5.2.

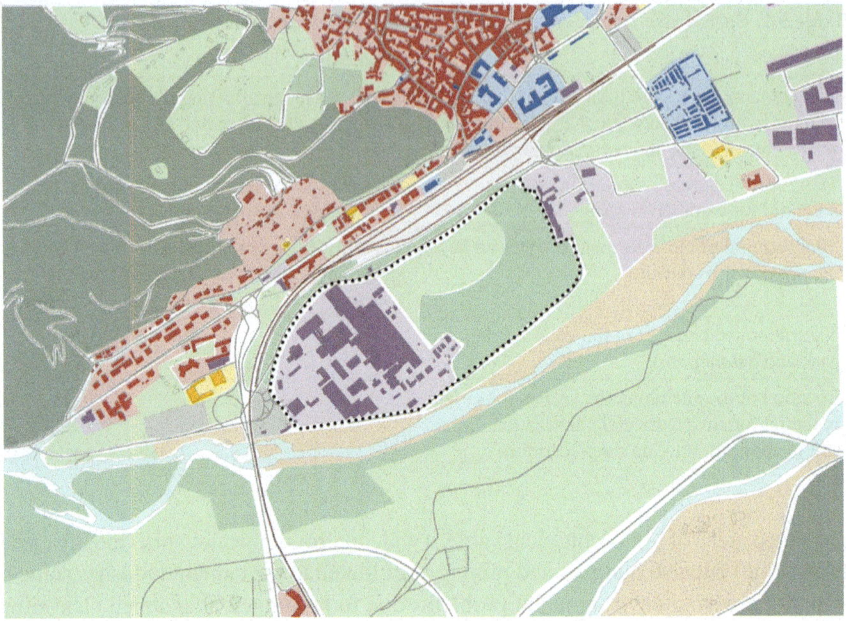

Fig. 5.3 Spatial assessment, settlements map. © Politecnico di Milano/Interreg project *trAILs*

Table 5.2 Spatial assessment, settlements-related potential and problems

Potentials	Problems
The site has a high potential for urban functions, thanks to its location in a flat area at the beginning of the valleys	The pilot area is an urbanized wedge inside a rich environmental system
Proximity of the pilot area to the historic settlement system	Disconnection of the site from the city core (road and railway)
High archeological potential due to the almost certain presence of an ancient roman cemetery	Separation of the site from the river due the to the wall and constructed embankment along the boundary of the area
Proximity to Cuneo (Borgo San Dalmazzo's helmets represent functional urban area of Cuneo)	

5.2.3 Mobility

The site lies at the crossroads of three valleys, and especially it is flanked by the SS20 road connecting Cuneo and Nice (France). Regarding public transport, the railway station of Borgo San Dalmazzo is located right next to the site, though accessible only on the northern side of the railway, and it is served by regional trains. Several bus lines are also passing nearby. The strategic location of the site in relation to public

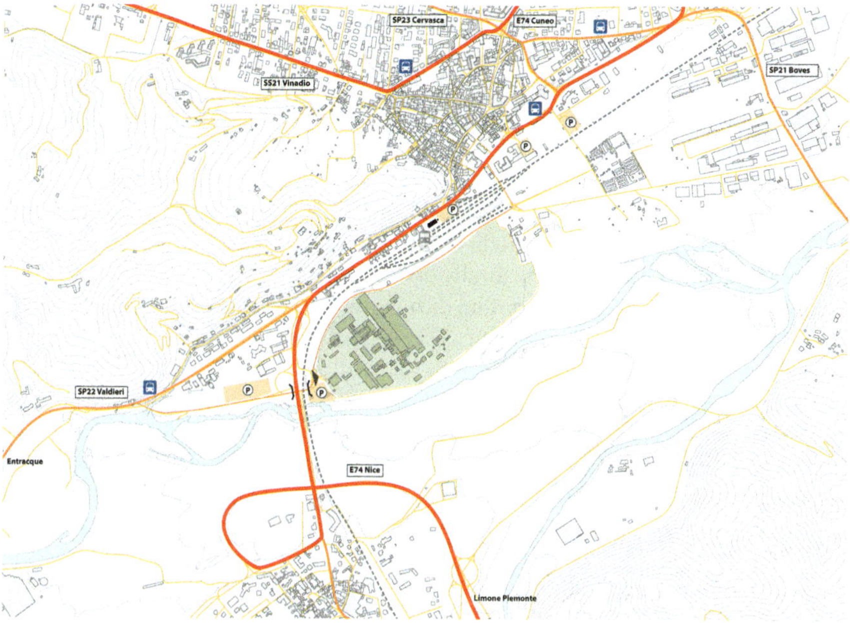

Fig. 5.4 Spatial assessment, mobility map. © Politecnico di Milano/Interreg project *trAILs*

Table 5.3 Spatial assessment, mobility-related potential and problems

Potentials	Problems
Proximity to the railway station, high potential for regional-wide connections by train	Many roads are not easily by some means of transport
High accessibility by road and bus lines, including seasonal public mobility for touristic purposes	The pilot area is not so connected with the surrounding areas
In general, very high accessibility of the site within the national, regional and local transport network (connection with France, Cuneo, Genova)	Shortage of local, pedestrian, bicycle and other recreational paths

transports and mobility networks describes constitutes a high potential, especially to overcome its actual physical isolation. The situation is portrayed in a specific map (Fig. 5.4) and described through its potentials and problems in Table 5.3.

In the view of the workshop, the spatial assessment was then integrated by the other thematic assessments carried on by the scientific partners, according to their own specialized expertise. The results are here briefly mentioned with reference to the thematic focus.

Socio-economic analysis[3] *(UNIVR)*: the local community is rather cohesive and heterogeneous in its composition but shows a low sense of belonging to the territory due to 'delocalized' networks and social relations. The labor market is dynamic and the employment rates in Borgo San Dalmazzo are above the average of Piemonte region. This makes the closure of the Italcementi site not really impacting on the socio-economic status.

Ecological analysis[4] *(TUM)*: the site is located close to an almost pristine, dynamic Alpine river ecosystem, with a high wilderness potential as habitat and migratory corridor. The site is however structurally disconnected from the river area and should benefit from a better integration within the landscape, in terms of ecological connectivity but also flooding management. The unbuilt area within the site has a high potential as habitat for many species.

Policy analysis[5] *(UL)*: at the regional and local level, brownfield sites are not mentioned as key challenges in landscape and spatial planning. At the same time, much effort is oriented towards defining strategies for sustainable territorial development and socio-economic growth, which can actually profit from a better recognition of brownfield redevelopment as a local priority.

5.3 Test-Design Workshop

Preliminary activities

The responsible project partner (POLIMI) involved a team of professional facilitators[6] to prepare and run the workshop. With their support, the workshop agenda was defined (three days, of which the second one with stakeholders) and then shared with the partners a couple of weeks before the event. The agenda included a side document summarizing in ten points the key workshop features and activities (purpose, duration, inputs, management, tables arrangement, working sessions, rules for group roundtables and plenary meetings, reflecting session, feedback rounds).

As design-based input material, useful to trigger the discussion around the roundtables, POLIMI and the other research partners prepared 15 thematic cards based on the results of the assessments (5 cards focusing on environmental topics, 3 on policy, 3 on territorial development, 4 on socio-economic issues). Furthermore, POLIMI and TUM prepared each a booklet presenting the design proposal for the site, as result from the teaching studios held by the two universities. To help reading and understanding each design proposal, as well as to compare them as a whole and through their composing parts, POLIMI designed and provided a comparative matrix (Fig. 5.5), a key tool to break down the complexity of the design in essential layers.

[3] Cf. *trAILs* Deliverables D.T2.2.4 (Veronesi et al. 2019) and D.T2.2.5 (Polin et al. 2019).

[4] Cf. *trAILs* Deliverable D.T2.2.6 (Strobl et al. 2019).

[5] Cf. *trAILs* Deliverable D.T2.2.2 (Krošelj et al. 2019).

[6] The team was led by Davide Fortini, an architect and expert in participatory planning with a specific focus on urban policies and environmental projects.

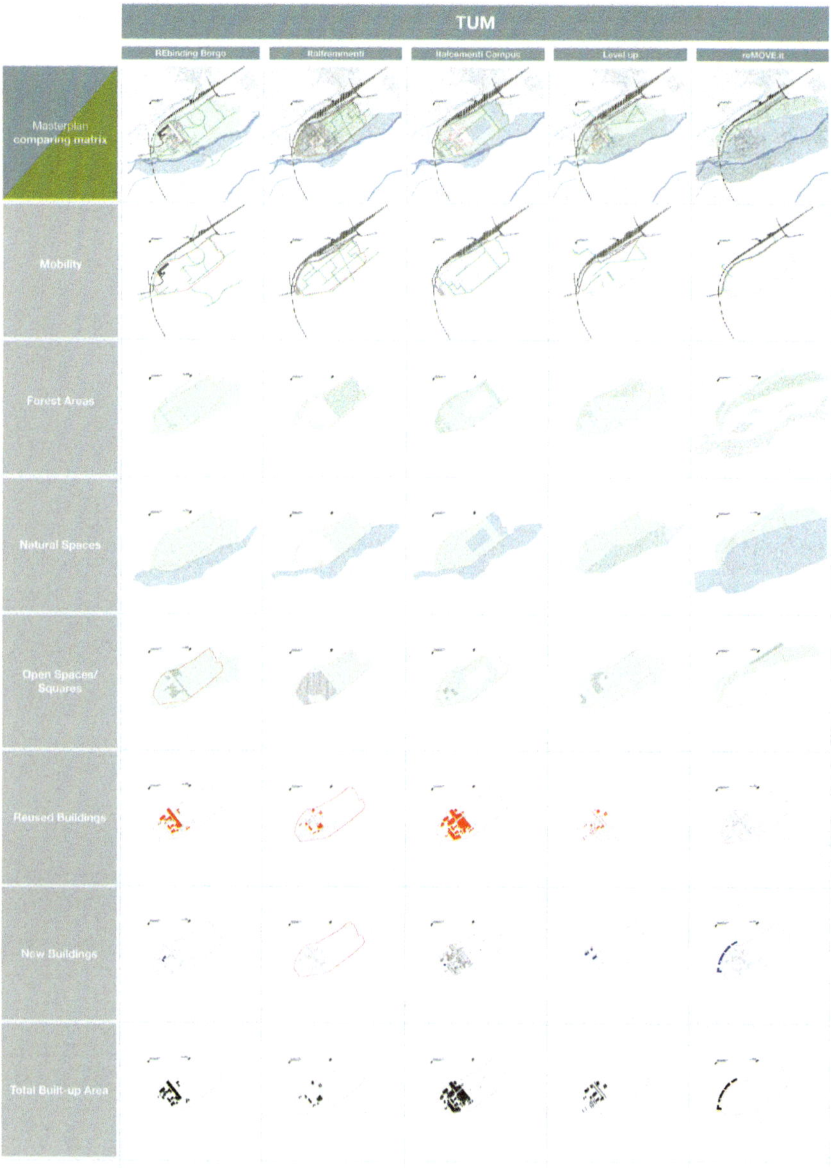

Fig. 5.5 The comparative matrix for reviewing and selecting the design proposals

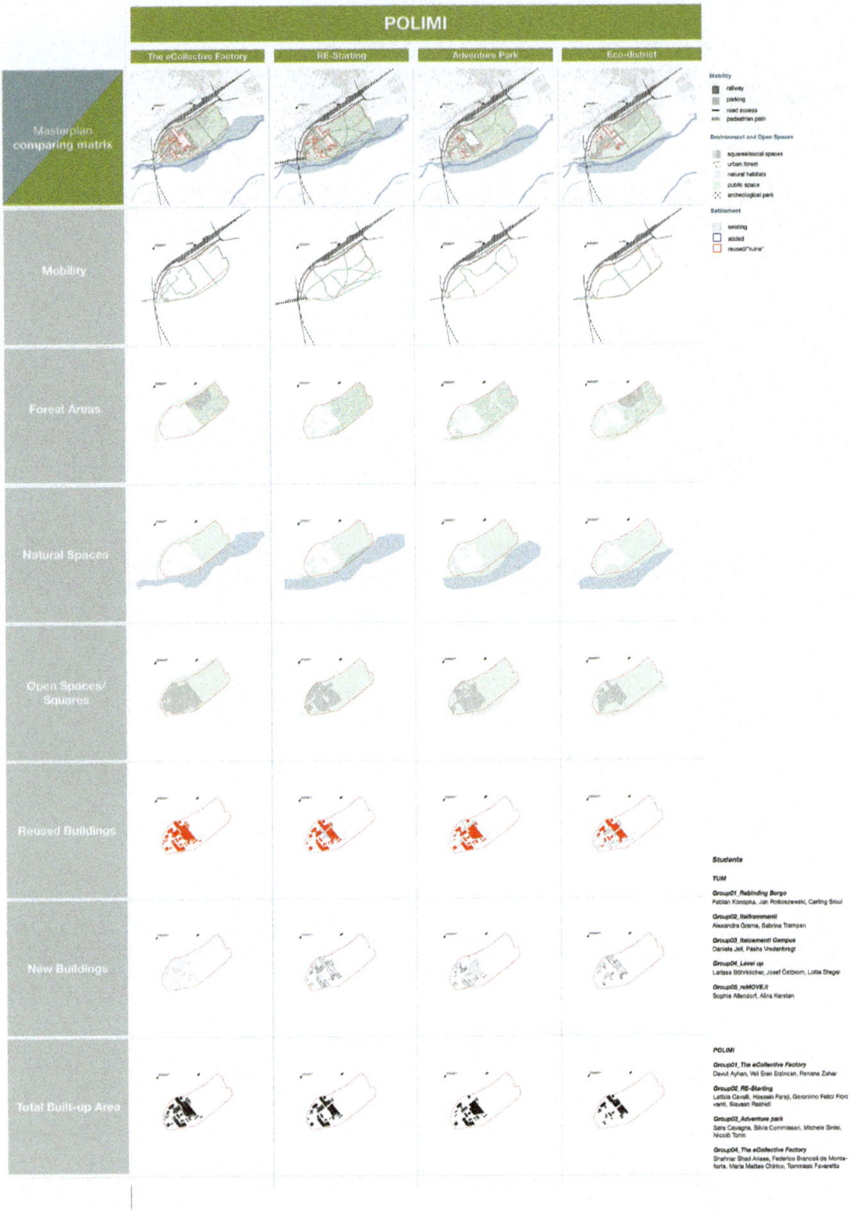

Fig. 5.5 (continued)

In terms of organization, the agreed workshop format was oriented towards setting three thematic tables (i.e. working groups) focusing on specific topics inspired by the cards (green table on environment and landscape; blue table on planning and policies; red table on economy and society). The arrangement of the working groups around each table had to be coordinated through a direct selection by the participating stakeholders.

A detailed list of stakeholders was previously drafted, together with the regional partner LAMORO. Around 50 among stakeholders and observers were contacted in advance, aiming at involving as more target groups as possible: municipality (mayor, local council, head of administrative/planning sectors), representatives of other neighboring municipalities, representatives of major cultural, tourist, environmentalist association, regional development agencies. Working materials to be arranged on the tables were provided in advance.

Workshop settings

The guiding principle of the workshop was set to guarantee the right to be listened even before the right to speak, to include everyone's opinion and learn from each other. Before the start of the roundtable session, a short video produced by UNIVR[7] was played to be seen by all the participants. The video described the everyday life in Borgo San Dalmazzo, focusing on issues such local identity, the impact of deindustrialization, the current living conditions and the prospects or expectations for the future. The video helped the present stakeholder to get engaged in the workshop aims and activities. Furthermore, TUM and POLIMI teams presented to the audience the design proposal (accompanied by the comparative matrix), as well as a poster illustrating the results of the thematic assessments (policy analysis, territorial and spatial analysis, socioeconomic analysis, ecological assessment)—assembled by the project partners under the coordination of UL.

After these presentations, greetings and registration, the participating stakeholders[8] (21 confirmed, 19 present) got acquainted with the thematic cards, explained by the facilitators, and eventually asked for adding new ones. The stakeholders then 'voted' those cards they considered the most significant for the site/territory and took a seat around the related thematic table. In this 'selection' procedure, the research partners and the observers supported the stakeholders with clarifications and integrations. Once the table were formed, the discussion could start, led by a facilitator.

[7] In cooperation with INTWIG communication and media agency.

[8] The participating stakeholders were: Ivo Riccardi (ARPA Cuneo), Claudio Re (Commercial consortium "Chiocciola di Borgo San Dalmazzo"), Daniela Risso (Legambiente Cuneo), Livio Quaranta (Azienda Cuneese Acque), Luca Gottero (River Park "Gesso Valley and Stura"), Gianpaolo Beretta (Mayor of Borgo San Dalmazzo), Giacomo Gaiotti (Mayor of Valdieri), Francesco Giraudo (municipality of Entraque), Guido Baschenis (Piedmont Region—Territory and Landscape Planning / observer), Patrizia Dalmasso (CNA Cuneo), Francesca Dalmasso (Confagricoltura Cuneo), Giuseppe Aimo (Confindustria Cuneo), Rita Piacenza (Confindustria Cuneo), Massimiliano Campana (CISL Cuneo), Vincenzo Battaglia (CISL Cuneo), Marco Borroni (cultural association), Elma Schena (CRC Foundation), Walter Cesana (citizen), Massimo Longhi (Touristic Office of the Province of Cuneo).

The same was also responsible to draft a sketchy poster including all the aspects emerged during the discussion (Fig. 5.6). Each table was provided with the same working materials: a poster containing thematic assessments, a blank map of the site, booklets of student's design, matrix.

After the table discussions, the working groups reunited in a plenary session. Here, each facilitator described the drafted poster of his/her respective table and then, through a group discussion, the key objectives and strategies for the site emerging from the tables and shared among the group were gradually identified. This phase was also supported by a visual designer, who 'reported' the discussion through artistic/pictorial design, aiming at generating an inspiring and visionary suggestion—for the participants but also for the broad local community. In the post-workshop phase, three different development paths were outlined, thanks to the useful and readable preliminary results obtained during the activities, and from these a key map, a synthesis of the commonly identified principles for the site transformation. The key map was then shared with the workshop participants (experts, communities, stakeholders) to get feedbacks on the actual understanding and accuracy with respect to the previous discussion. The reviewed key map constituted the basis for the construction of design recommendations useful for orienting the future of the site.

To allow a clearer and homogeneous interpretation of the results from each of the thematic tables, the essential content of the discussion is presented in the following three paragraph through a set of key questions and topics, with which each table had to confront.

Roundtable 1 (environmental issues)

Group: 5 stakeholders, 8 research partners, 1 moderator/interpreter.

Built heritage within the site: conservation or recovery/reuse, which suitability and criteria?

- Balance the preservation of the historical memory of the site with regeneration needs (even in terms of extensive renaturalization).
- Reuse existing buildings by assigning new functions related to tourism, especially in connection with the regional offer.
- Consider the operational difficulties and intervention costs of conservation, as well include an assessment of architectural quality and sense of belonging of the local community (to validate conservation and reuse).

A transformation oriented to young generations (functions, uses, involvement, positive outcomes).

- The area could become the positive response to the request for employment of young people (new functions, touristic uses).

Enhancing the physical/visual permeability of the site while keeping alive the memory of the past, how to combine?

- The cement factory is a strong physical fracture that also visually interrupts the town-river connection; at the same time the presence of this industrial activity has allowed the development of transport connections (roads and railway), opening the town to the context; a first solution to eliminate the fracture could be to demolish the walls along the perimeter of the site and provide useful devices (barriers) to prevent flooding from the river Gesso. Some industrial buildings represent a significant landmark and should be maintained (e.g. the factory chimneys).

How to link the site transformation with the peculiarities and/or critical factors of the mountain landscape?

- The territory of Borgo San Dalmazzo and the surrounding valleys are rich in vegetation coverage and biodiversity. The Gesso river park, which flanks the site, could help to safeguard the river shores from urban expansion and soil sealing, and transformation should sustain this potential.
- River as an important element giving identity to the place/city (local community/economy), integrating well the flood risks.

Environmental remediation and reclamation: how to incorporate the water element in the process?

- The Gesso river is a valuable natural resource to be enhanced and integrated into the project; however, it is necessary to restore and secure the surrounding areas in case of flooding events.
- Promote non-invasive and nature-based interventions to restore the riverbed while limiting the impact of flooding.
- Solve water pollution caused by industrial processes.

Enhancing the strategic position of the site with respect to transport routes (road and railway accessibility).

- Promote the creation of a multimodal transportation hub integrating railway, roads and slow mobility (cycling and trekking paths along the river and the valley system). The railway line must be kept in operation and strengthened due to the potential tourist vocation of the area.

Foster inclusive decision-making processes (active involvement of the site owners, harmonization of multi-scale planning schemes and guidelines).

- Participatory planning is the only way to achieve community-oriented solutions, not only for Borgo San Dalmazzo but especially for the valley communities in the surroundings.
- Consider in advance the operational difficulties and costs for all the planned interventions, as well as carefully assess the spatial-architectural quality and the social identity/demand coming from the local community.

How to access financing as support for the site transformation?

- This roundtable did not consider this particular issue, except as for emphasizing the need for a real participatory planning approach (in matter of financing too).

Roundtable 2 (governance and policy issues)

Group: 4 stakeholders, 4 research partners, 1 moderator/interpreter.

Built heritage within the site: conservation or recovery/reuse, which suitability and criteria?

- Balance the preservation of the memory of the site and its physical and functional regeneration.
- Reuse of buildings with a high technical/constructive as well as artistic value.
- Consider operational difficulties and reclamation costs.

A transformation oriented to young generations (functions, uses, involvement, positive outcomes).

- The site transformation can provide a positive response to face the migration issues; besides, it is very important to promote professional and vocational training activities in local schools, aiming at responding to the local job market as well as to improve the offer/demand. It is necessary to develop new professional skills that can deal with the potential of mountain resources/spaces such as forests or the former quarrying sites.

Enhancing the physical/visual permeability of the site while keeping alive the memory of the past, how to combine?

- Promote a transformation process able to mix recovery actions (for example on significative and identity-bearing buildings and structures) with demolition activities.

How to link the site transformation with the peculiarities and/or critical factors of the mountain landscape?

- Rediscover and enhance local expertise in the maintenance of forests (thinning and cutting) and the regeneration of slope terraces for agricultural use (expert craftsmen with the knowledge of traditional techniques of dry-stone walls).

Environmental remediation and reclamation: how to incorporate the water element in the process?

- Solve water pollution caused by industrial processes.
- Integrated water-soil remediation, considering the agricultural and residential surroundings.
- Need to further investigate the issues highlighted in the Regional territorial and landscape plans.

Enhancing the strategic position of the site with respect to transport routes (road and railway accessibility).

- This roundtable did not consider this particular issue, except as for highlighting the strategic position of the site.

Foster inclusive decision-making processes (active involvement of the site owners, harmonization of multi-scale planning schemes and guidelines).

- Strengthen the relationship between political administration and university/research, with focus on the site transformation.
- Promote sustainable territorial development projects in cooperation with research institutions, to increase the capability of local authorities in defining and promoting feasible and innovative strategies.
- Activate a decision-making process based on community involvement.
- Joint efforts towards achieving a unified vision within the territorial governance processes shared between the municipality, province and region (formally, an *Accordo di Programma*).
- Legally implement a land use change from "Productive area" to "Transformation area".

How to access financing as support for the site transformation?

- Two possible scenarios: (a) the private (Italcementi company) finances the public, which becomes a partner in promoting the redevelopment of the site; (b) the private sells the site to the public and new investors enter the process: in this case the public must be ready with a feasibility plan.
- Promote temporary uses while waiting for the long-term transformation.

Roundtable 3 (socio-economic issues)

Group: 10 stakeholders, 3 research partners, 1 moderator/interpreter.

Built heritage within the site: conservation or recovery/reuse, which suitability and criteria?

- Preservation of some structures of historical and heritage value and demolition of less significant or unsafe buildings.
- Reuse of existing buildings and structures for tourism, economic and cultural activities linked to the territory, as well as for enhancing the history of the site.

A transformation oriented to young generations (functions, uses, involvement, positive outcomes).

- Establish on the site new activities and job opportunities for young people: productive and economic development connected to tourism, as well as promotion of local skills and products (food and wine and crafts).
- Establish a new university-related function ("Alpine College/University Campus"), i.e. a place of educational training connected with the territory.

Enhancing the physical/visual permeability of the site while keeping alive the memory of the past, how to combine?

- Preservation of the chimneys and ovens as a museum/exhibition place to remind the history of cement production.

How to link the site transformation with the peculiarities and/or critical factors of the mountain landscape?

- Link the environmental sustainability of the site regeneration to the territorial characteristics and potentials.
- Promotion of local skills and zero-km food production (enogastronomic and handcrafted activities).
- Promotion of regional-based alpine tourism by enhancing the local spa offer (Vinadio and Valdieri bath towns, to which add Borgo San Dalmazzo in the future) and outdoor sports along the river.
- New educational and training center strictly linked to the territory.
- Centre for historical memory of the site, buildings as containers of tourist, economic and cultural activities linked to the territory.

Environmental remediation and reclamation: how to incorporate the water element in the process?

- Promote a comprehensive site redevelopment, by integrating built spaces and green areas connected to a river park (where rafting and canoeing are practiced).

Enhancing the strategic position of the site with respect to transport routes (road and railway accessibility).

- The transport and traffic issues must be considered as key element in the site transformation, since the latter is located in a strategic position between communication routes and near to cycle paths, as well as to rail networks.
- Strengthen road connections and cycle-pedestrian touristic routes, such as the so-called "Via del sale" connecting Piedmont, Liguria and France.
- Promote the creation of a naturalistic route along the river.

Foster inclusive decision-making processes (active involvement of the site owners, harmonization of multi-scale planning schemes and guidelines).

- Set up a permanent territorial discussion table that involves as many stakeholders as possible in monitoring the overall project and individual choices.

How to access financing as support for the site transformation?

- Consider the different funding opportunities and how to access European funds.
- Promote tax incentives based on energy efficiency.

5.4 Planning Recommendations

In Borgo San Dalmazzo the planning recommendations could be easily drafted at the end of the workshop, starting from the outcomes of the roundtable discussion. The preliminary results, in fact, were sketched planning schemes highlighting, for each of the table, mutually agreed planning issues on different spaces and themes. In the post-workshop phase, the effort consisted in outlining the final draft of the planning recommendations, inclusive of all the emerged key element as well as capable of synthesizing the outcomes. The results of this process consisted in three alternative development paths, ranging from the 'softer' transformation (start-up/temporary activation) to the most radical ones ('back-to-city' and 'back-to-nature'), as well as in a key map, an effective planning tool describing the essential spaces and interventions on which the site regeneration should be based. For each development path, the key actions related to the structuring issues for the site transformation—ecological network and public open spaces (including potentially relating built-up ones)—are described with text and an attached illustrative plan.

Start-up/temporary activation

In view of a comprehensive transformation of the former industrial area, the reactivation of some of its parts might help, as a priority intervention, to trigger the whole regeneration process. The 'start-up' path envisages specific actions aimed at preparing the area, such as the preservation and reuse of the most prominent built ensemble (former rotating kilns halls and related chimneys), as well as some smaller structures near the former site entrance. The site reactivation involves the identification and development of a minimal ecological and mobility/open spaces network, in order to connect the activated parts of the site and the outer areas (Table 5.4; Fig. 5.7).

Table 5.4 Start-up scenario, key issues

Ecological network	*Landscape-environmental recovery of the Gesso riverbanks*: create new leisure facilities on the banks, new pathways, landscaping
	Gesso river: manage the flooding risk on the site by creating a buffer zone between the river and the inner portions of the site, in order to minimize flooding impacts on and beyond the site (towards the town)
	Increasing of vegetation cover: enhance the semi-natural look of the site by expanding and improving the actual vegetation cover
	Chimneys area: clarify the level of contamination and consider measures for soil remediation, if necessary
	Green pathways: create a cycle/pedestrian path on the southern border of the site, as to connect the river landscape with the site and the urban area around it
	Ecological connections: connect the green spaces within the site with the ecological systems in the surroundings
Public open spaces	*Public open spaces and pathway network*: connect the activated parts of the site and enhance the general livability of the area
	Industrial heritage recovery and enhancement (rotating kilns building, chimneys): preserve the historical built structures as site landmarks, while planning temporary or permanent uses inside and around
	Refurbishment of the entrance area: renew the former site entrance (open spaces and buildings) as to show the transformation ongoing
	Selective demolition of the site perimetral walls: removal of the concrete walls around the site; along the river side, the removal should include nature-based solutions to prevent eventual flooding

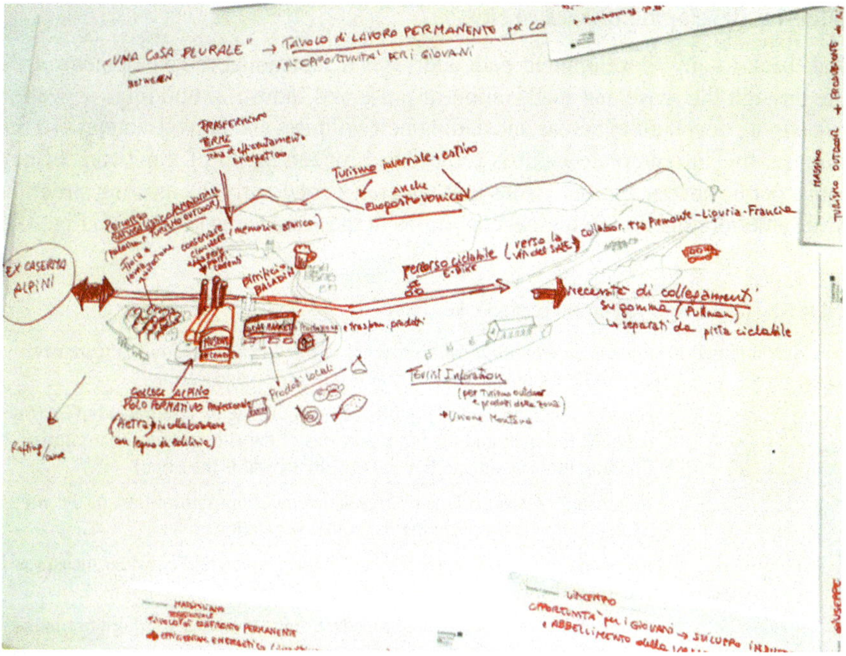

Fig. 5.6 One of the sketched posters resulting from a thematic working table. © Roberta Marchesi

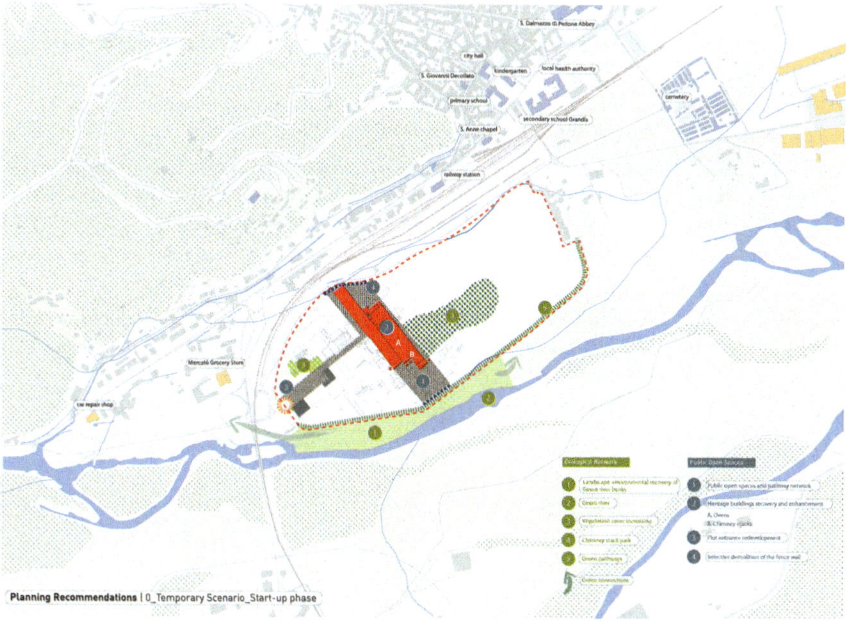

Fig. 5.7 Start-up scenario. © Politecnico di Milano/Interreg project *trAILs*

Intensive development/ *back to city*

The 'back-to-city' development path addresses a substantial redevelopment of the site through the reuse and reactivation of preserved industrial buildings as well as by defining new building areas and adding new volumes and built structures. To this purpose, the site internal layout is refurbished by setting-up of a network of new paths, public spaces and servitudes that integrates coherently the existing structure, while ameliorating the ecological conditions of the whole area (Table 5.5; Fig. 5.8).

Table 5.5 Intensive development scenario, key issues

Ecological network	*Landscape-environmental recovery of the Gesso riverbanks*: create new leisure facilities on the banks, new pathways, landscaping
	Gesso river: manage the flooding risk on the site by creating a buffer zone between the river and the inner portions of the site, in order to minimize flooding impacts on and beyond the site (towards the town)
	Increasing of vegetation cover: enhance the semi-natural look of the site by expanding and improving the actual vegetation cover
	Chimneys area: clarify the level of contamination and consider measures for soil remediation, if necessary
	Green pathways: create a cycle/pedestrian path on the southern border of the site, as to connect the river landscape with the site and the urban area around it
	Archeological Park: creation of a public green area within the site perimeter, to valorize the ancient Roman heritage there present
	Ecological connections: connect the green spaces within the site with the ecological systems in the surroundings
Public open spaces	*Public open spaces and pathway network*: connect the activated parts of the site and enhance the general livability of the area
	Industrial heritage recovery and enhancement (rotating kilns building, chimneys): preserve the historical built structures as site landmarks, while planning temporary or permanent uses inside and around
	Refurbishment of the entrance area: renew the former site entrance (open spaces and buildings) as to show the transformation ongoing
	New buildings: to host new functions for the site future development, new buildings should be realized between the former industrial buildings and the site entrance

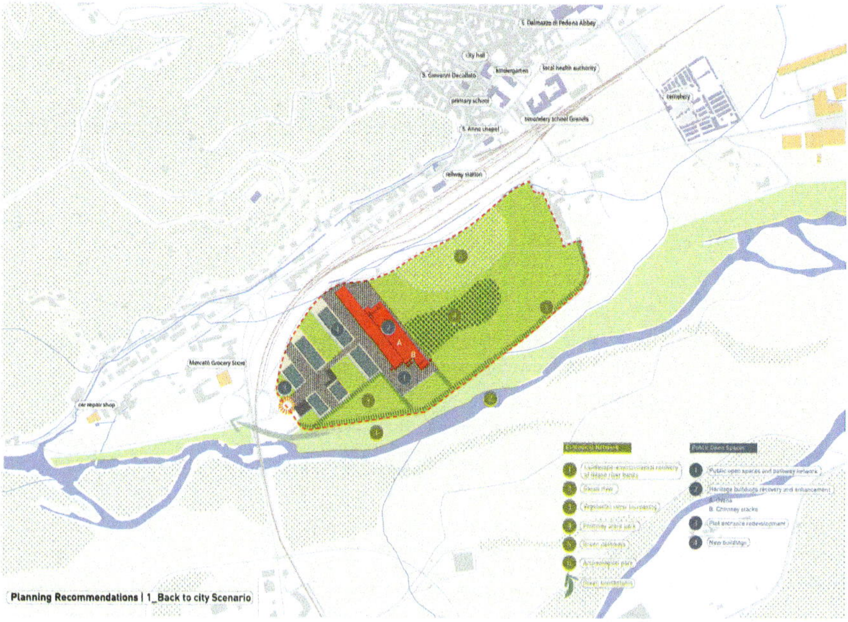

Fig. 5.8 Intensive development scenario. © Politecnico di Milano/Interreg project *trAILs*

Extensive development/ *back to nature*

The 'back-to-nature' development path deals with the foreseeable perspective of a progressive underusage of the site. The conservation of just a few relevant built structures (identified as industrial heritage) coexists with an extensive environmental regeneration of the former industrial area, in coherence with the surrounding landscape. The renaturation of the site involves three main phases and relative regenerative landscapes: (1) sparsely vegetated landscape with spontaneous succession, (2) grassland-dominated landscape, (3) forest-dominated landscape (Table 5.6; Fig. 5.9).

Key Map

As shared between the three development paths, and preliminary to their future implementation, the key map integrates and illustrates the commonly agreed issues as identified in the test-design workshop. These include the essential network of public spaces, the environmental system, and relevant ecological spaces as well as the most relevant buildings and structures to be preserved and reused. By displaying the basic structure for the future development of the site, the key map is a valuable tool in the hands of the community, a starting point for any possible transformation. The minimum requirements for the site transformation and regeneration, no matter what will be the scenario that the community implement, are listed here below (Table 5.7; Fig. 5.10).

Table 5.6 Extensive development scenario, key issues

Ecological network	*Landscape-environmental recovery of the Gesso riverbanks*: create new leisure facilities on the banks, new pathways, landscaping
	Gesso river: manage the flooding risk on the site by creating a buffer zone between the river and the inner portions of the site, in order to minimize flooding impacts on and beyond the site (towards the town)
	Increasing of vegetation cover: enhance the semi-natural look of the site by expanding and improving the actual vegetation cover
	Chimneys area: clarify the level of contamination and consider measures for soil remediation, if necessary
	Green pathways: create a cycle/pedestrian path on the southern border of the site, as to connect the river landscape with the site and the urban area around it
	Archeological Park: creation of a public green area within the site perimeter, to valorize the ancient Roman heritage there present
	Ecological connections: connect the green spaces within the site with the ecological systems in the surroundings
Public open spaces	*Public open spaces and pathway network*: connect the activated parts of the site and enhance the general livability of the area
	Industrial heritage recovery and enhancement (rotating kilns building, chimneys): preserve the historical built structures as site landmarks, while planning temporary or permanent uses inside and around
	Refurbishment of the entrance area: renew the former site entrance (open spaces and buildings) as to show the transformation ongoing

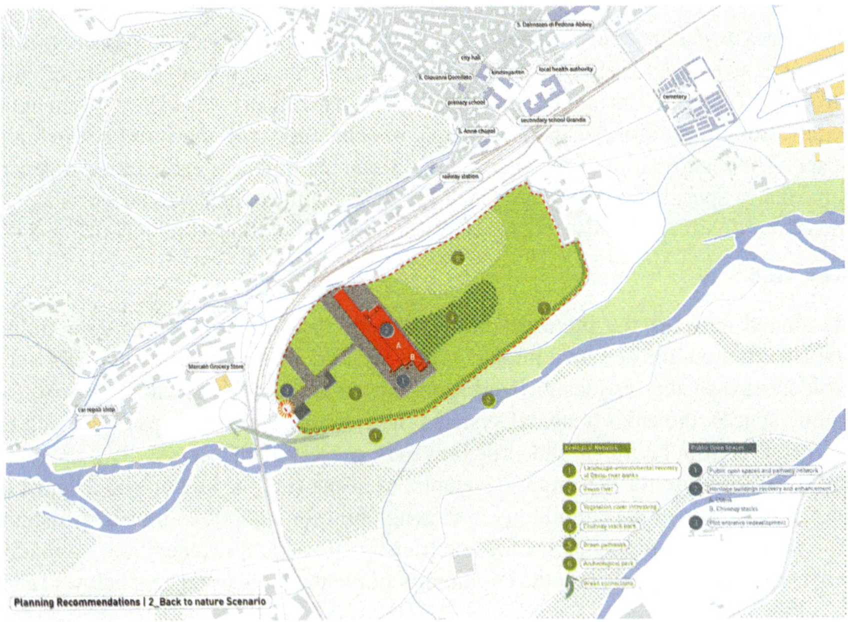

Fig. 5.9 Extensive development scenario. © Politecnico di Milano/Interreg project *trAILs*

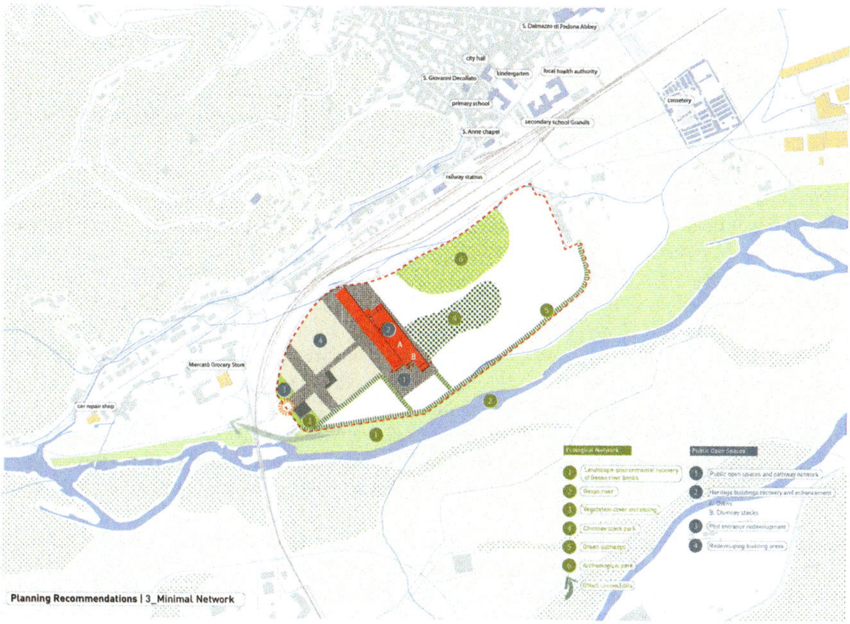

Fig. 5.10 Key map. © Politecnico di Milano/Interreg project *trAILs*

Table 5.7 Key map issues

Landscape-environmental recovery of the Gesso riverbanks: create new leisure facilities on the banks, new pathways, landscaping
Gesso river: manage the flooding risk on the site by creating a buffer zone between the river and the inner portions of the site, in order to minimize flooding impacts on and beyond the site (towards the town)
Increasing of vegetation cover: enhance the semi-natural look of the site by expanding and improving the actual vegetation cover
Chimneys area: clarify the level of contamination and consider measures for soil remediation, if necessary
Green pathways: create a cycle/pedestrian path on the southern border of the site, as to connect the river landscape with the site and the urban area around it
Archeological Park: creation of a public green area within the site perimeter, to valorize the ancient Roman heritage there present
Ecological connections: connect the green spaces within the site with the ecological systems in the surroundings
Public open spaces and pathway network: connect the activated parts of the site and enhance the general livability of the area
Industrial heritage recovery and enhancement (rotating kilns building, chimneys): preserve the historical built structures as site landmarks, while planning temporary or permanent uses inside and around
Refurbishment of the entrance area: renew the former site entrance (open spaces and buildings) as to show the transformation ongoing
New building zones: identify appropriate building zones for the site future development, to be concentrated between the former industrial buildings and the site entrance

References

Fava U, Abluton S, Curato V (2018) EU Interreg project trAILs—Deliverable D.T1.3.2.—Pilot Profile Borgo San Dalmazzo. Available at: https://www.alpine-space.org/projects/trails/pilots/_results/wp-t1/pilot_profile_it_final.pdf

Krošelj M, Pipan T, Marot N et al (2019) EU Interreg project trAILs—Deliverable D.T2.2.2.—Policy assessment report Borgo San Dalmazzo. Available at: https://www.alpine-space.org/projects/trails/pilots/_results/wp-t2/trails_d_t2_2_2_policy_regional_report_italy_lamoro.pdf

Polin V, Prete V, Fava U et al (2019) EU Interreg project trAILs—Deliverable D.T2.2.5.—Economic assessment report Borgo San Dalmazzo. Available at: https://www.alpine-space.org/projects/trails/pilots/_results/wp-t2/trails_d_t2_2_5_economic_assessment-report_italy_univre.pdf

Solero E, Vitillo P, Galuzzi P et al (2019) EU Interreg project trAILs—Deliverable D.T2.2.3.—Spatial planning and landscape assessment report Borgo San Dalmazzo. Available at: https://www.alpine-space.org/projects/trails/pilots/_results/wp-t2/trails_d_t2_2_3_spatial_assessment_report_italy_final_tuw_polimi.pdf

Strobl K, Bär K, Fava U et al (2019) EU Interreg project trAILs—Deliverable D.T2.2.6.—Environmental assessment report Borgo San Dalmazzo. Available at: https://www.alpine-space.org/projects/trails/pilots/_results/wp-t2/trails_d_t2_2_6_environmental_assessment_report_italy_tumroek.pdf

Veronesi L, Migliorati L, Fava U et al (2019) EU Interreg project trAILs—Deliverable D.T2.2.4.—Socio-demographic assessment report Borgo San Dalmazzo. Available at: https://www.alpine-space.org/projects/trails/pilots/_results/wp-t2/trails_d_t2_2_4_social_assessment-report_italy_univrhs.pdf

Chapter 6
L'Argentière-la-Bessée, France

The chapter presents the French pilot site through the activities and achieved results in the framework of the trAILs project. A brief overview of both the regional context and the brownfield site is first provided, describing the main features and transformation challenges. Secondly, the identified potentials and problems are listed, as emerged from to the spatial assessment and the site visit. The chapter describes then the test-design phase, focusing on the workshop setting and methodological aspects and procedures. Finally, the planning recommendations for the site are presented, as outlined during and after the workshop.

6.1 Regional and Site Overview[1]

The Durance river and its valley are an ancient transalpine route connecting the Marseilles region and Provence with the northern French Alpine regions (Isère, Savoie), as well as with the Italian Piedmont towards east. The upper Durance valley extends for nearly 50 km between the historic centers of Embrun (6857 inhabitants) and Briançon (16,355 inhabitants), providing access, through high mountain passes, to the routes to Turin and Grenoble. Due to the remote geographical setting, the region is today hardly accessible on the long distance—the nearest airports are Turin and Grenoble, 2.5 h away, and Marseille, 3 h. The department of Hautes-Alpes, in which the pilot region is entirely located, is a sparsely populated rural area with 141,000 inhabitants and a density of 23 inhabitants/km^2. Around 80% of the land is covered by forests, mountain pastures and bare rocks. The upper Durance valley as a cultural region is home to 40,000 inhabitants distributed across 42 municipalities, often comprising several little and dispersed mountain hamlets. Despite its location, the region has registered a constant demographic growth since the 1980s, mostly due to amenity migration by retired people or young families. The economy is strongly

[1] Information retrieved and adapted from *trAILs* Deliverable D.T1.3.2 (Kleitz 2018).

M. Modica and E. Solero, *Brownfield Transformation in Fragile Territories*, PoliMI SpringerBriefs, https://doi.org/10.1007/978-3-031-09083-7_6

tied to year-round tourism, thanks to the scenic mountain landscapes and the mild Mediterranean climate, with several ski resorts and vast protected natural sites (Ecrins National Park and Queyras Regional Natural Park). In the past, the region never experienced a significant industrial development as in the northern French Alps, except for a few remarkable initiatives dating back to the early twentieth century and related to hydroelectric energy production.

One of those experiences led to the establishment of the Péchiney aluminum smelter in L'Argentière-la-Bessée. The first hydroelectric power station was built in 1907, followed by aluminum production in 1910. The site gradually expanded until the 1960s, when it reached its peak of production while employing more than 1.500 workers. The economic difficulties experienced in the 1970s energy and industrial crisis led to the definitive closure of the plant in 1985. The core production site was partially refurbished in 1987–1988 to host a steel foundry (AFP), while the northern site (carbon black plant) was cleared and redeveloped as housing and commercial estate. After the failure and closure of AFP in 2012, part of the area was bought by the intermunicipal association of Pays des Ecrins (CCPE) and roughly reused by small craft activities and local services (ZA Les Sablonnières). In 2019, CCPE finally took over the rest of the area, actually derelict and unused (Fig. 6.1).

The whole site appears today as a patchwork of abandoned, underused or refurbished buildings and open spaces, with approximately 85% of the land built-up. A few heritage buildings, belonging to the original aluminum factory, are waiting to

Fig. 6.1 Overview of the Pechiney-AFP brownfield. © Marcello Modica

be repurposed—among those, the huge former hydropower station from 1907. The southern portion of the site, mostly unbuilt, is affected by subsoil contamination. Part of the contaminated area is occupied by a former slag dump (by-product waste), not yet reclaimed and spontaneously renatured, which lays close to a public green space and sport/leisure center. The site is rather well-connected, having a good road and railway accessibility (the station is located 500 m away) as well as available energy and water in the northern (used) part. Concerning its future development, the main challenge is related to pursuing a comprehensive and sustainable transformation, capable of linking the different sections of the site (abandoned and used, old and new, built and unbuilt) to the surrounding urban and ecological systems. The strategic location of the site and its size compared to the town of L'Argentière make its transformation a unique chance to rethink the whole urban structure, as well as to improve its appearance and/or attractiveness. In particular, the relationship between the site and the Durance river is crucial to this achievement. On the one hand, the site transformation should incorporate all the necessary risk-management measures to contain flooding (as the canalized river runs nearby). On the other hand, the site redevelopment should rely on the river as an historic element (hydropower heritage) as well as a space for sports, leisure and tourism. The public ownership of the whole area, after the recent takeover of the AFP site, represents a great advantage. On the economic side, the challenge is twofold: first, to find enough resources to complete the site decontamination, and secondly to develop the site in a way that the future activities will be able to generate long-term jobs for the community and the whole region.

6.2 Spatial Assessment[2]

The spatial assessment of the site and its context began with a site visit, which took place on 21–22 November 2019. In the site visit, organized by CAUE 84 and CAUE 05, the project partners were guided by local representatives and experts on the former Péchiney site, as well as on a second brownfield site (also belonged to Péchiney) located in the nearby town of La-Roche-de-Rame. The actual visit on the sites was conducted by foot and lasted for more than half a day, accompanied by very detailed explanations by the local experts on the site conditions, challenges and main issues. In the afternoon of the first day, then, the project team had the chance to deepen these issues by meeting the local and regional stakeholders (mayors of L'Argentière and La-Roche, representatives of the CCPE and PACA region) in a round table session. The stakeholders answered the questions of the project experts on different topics, such as the management of soil contamination, the urban development plans, the functional link between the two brownfields and the socioeconomic impact of transformation. This exploratory discussion was extremely useful to the partners to draft their thematic assessments. Concerning specifically the spatial

[2] Cf. *trAILs* Deliverable D.T2.2.3 (Solero et al. 2020).

assessment, the knowledge gained on-site during the visit and the roundtable with the stakeholders was then integrated, off-site, with the local planning documents provided by the regional partners. Different aspects were covered, from the regional accessibility to the local conditions of the site and its context. The resulting spatial assessment is presented below through its three main analytical frameworks: environment, settlements and mobility. Further analyses were also done (e.g. land property, built heritage, etc.) and used in the workshop.

6.2.1 Environment

The site and its immediate surroundings are characterized by buildings and paved, impermeable open spaces, with few or no significant environmental qualities (except for the southern edge, which is however partially contaminated). The context, and especially the mountain slopes and the left bank of the Durance, present a diverse environmental pattern with high ecological and landscape value. The situation is portrayed in a specific map (Fig. 6.2) and described through its potentials and problems in Table 6.1.

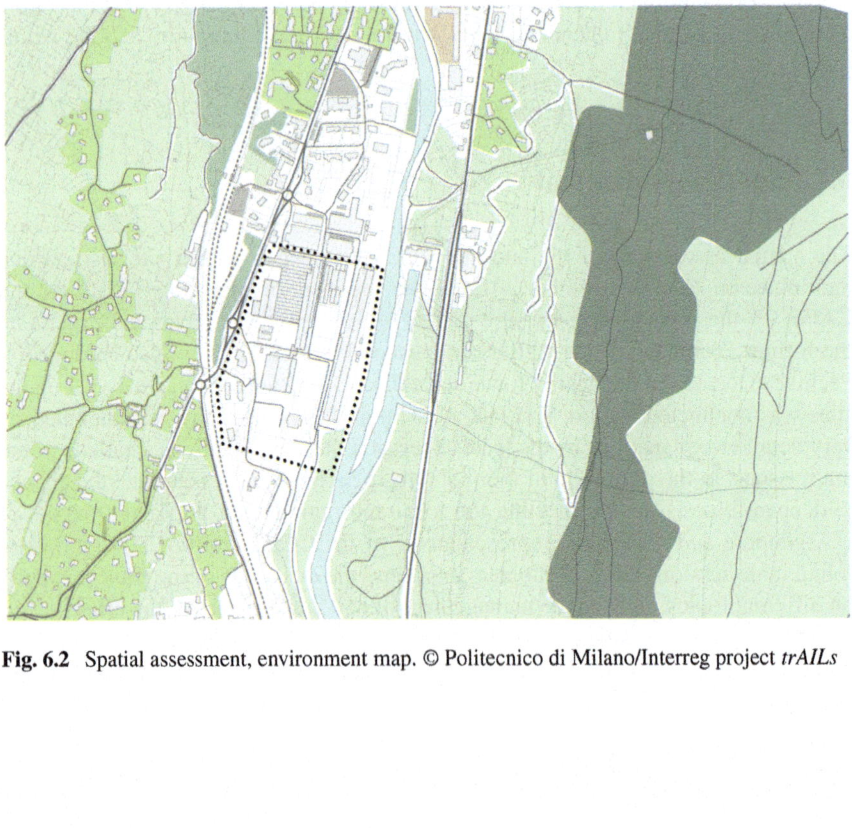

Fig. 6.2 Spatial assessment, environment map. © Politecnico di Milano/Interreg project *trAILs*

Table 6.1 Spatial assessment, environment-related potentials and problems

Potentials	Problems
Diverse landscape, with high potential in terms of biodiversity and ecological connections as well as for touristic and recreational purposes	Climate change affecting the water regime of the Durance, thus increased risk of natural hazards such as flooding for the site and the town too
Many natural or close-to-natural water bodies in the proximity of the site	Pollution in the southern edge of the site (former slag dump)
Natural protected areas in the vicinity (national park and nature park)	
Location of the site along the Durance, thus high potential for environmental valorization of the site in connection to the river corridor (valley-wide green infrastructure)	

6.2.2 *Settlement Structure*

The site is characterized by industrial buildings with a very large build footprint compared to the surrounding, standing on both sides of the Durance. When used, the buildings are mostly occupied by crafts and small production activities. The settlement structure of the surroundings is characterized by scattered residential areas (sprawled hamlets) on the slopes and compact urbanization in the valley floor (former Péchiney workers settlement, including most of the public facilities and

Fig. 6.3 Spatial assessment, settlements map. © Politecnico di Milano/Interreg project *trAILs*

Table 6.2 Spatial assessment, settlements-related potentials and problems

Potentials	Problems
The site is centrally located within L'Argentière-la-Bessée, within walking distance from main services and facilities	Oversize and mono-functionality of the existing buildings within the site
Good level of integration of the site within the existing urban structure, especially in the northern half	The site occupies the whole right bank of the Durance, in the valley floor, thus interrupting (disconnecting) the urban structure of L'Argentière
The site has a compact urban footprint, favorable to inner development and densification	High fragmentation of the land property within the site

commercial services). The situation is portrayed in a specific map (Fig. 6.3) and described through its potentials and problems in Table 6.2.

6.2.3 *Mobility*

Due to its central location, the site is well connected to the road and railway network, as well as to the local public transport network. The railway station of L'Argéntiere is within walking distance from the site's northern edge. Traffic congestion along the RN94 negatively affects the western portion of the site (AFP brownfield). The

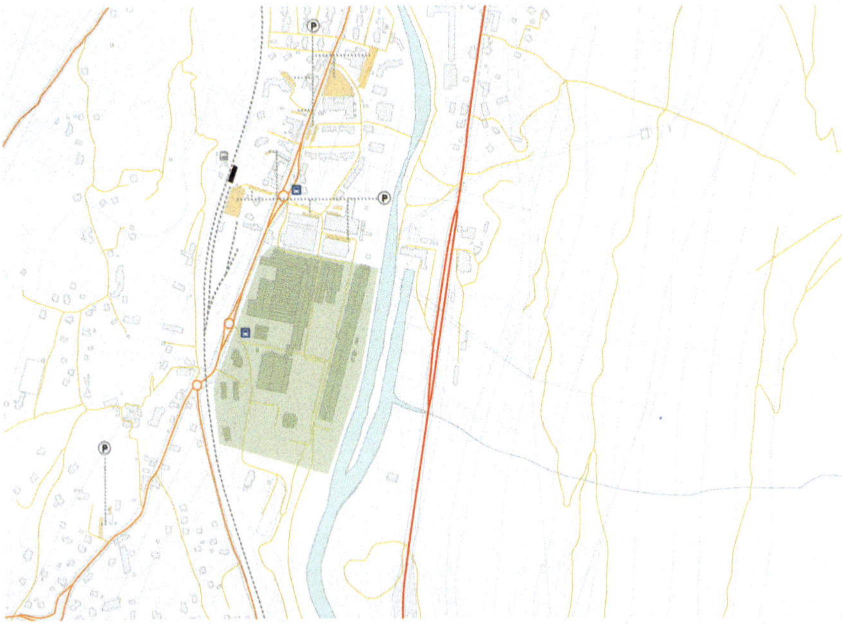

Fig. 6.4 Spatial assessment, mobility map. © Politecnico di Milano/Interreg project *trAILs*

Table 6.3 Spatial assessment, mobility-related potentials and problems

Potentials	Problems
Proximity to the railway station, high potential for regional-wide connections by train	Temporary RN94 traffic congestion nearby the site, especially in the winter season due to skiing tourism
High accessibility by road and bus lines, including seasonal public mobility for touristic purposes	Underdeveloped road network at the local and regional scale
In general, very high accessibility of the site within the regional and local transport network	Underused of the local railway line for passenger and freight transport (risk of line closure)

situation is portrayed in a specific map (Fig. 6.4) and described through its potentials and problems in Table 6.3.

In the view of the workshop, the spatial assessment was then integrated by the other thematic assessments carried on by the scientific partners, according to their own specialized expertise. The results are here briefly mentioned with reference to the thematic focus.

Socio-economic analysis[3] *(UNIVR)*: the local community is socially cohesive, though the economic system lacks opportunities for young generations, fostering outmigration. A participatory approach to the site's transformation is required, to empower the trust in local institutions and enhance the transparency of decision-making.

Ecological analysis[4] *(TUM)*: despite lacking significant natural ecologies, due to the mostly built-up surfaces, the site is a local hotspot for several flora and fauna species. Depollution of contaminated areas is recommended as well as the conservation and improvement of habitats.

Policy analysis[5] *(UL)*: at the regional and local level, it shows a lack of understanding of brownfields as an opportunity for sustainable urban and territorial regeneration. A long-term support for brownfield redevelopment is missing, while tools and incentives for the initial stages (reclamation, ownership) are existing and used.

---

[3] Cf. *trAILs* Deliverables D.T2.2.4 (Veronesi et al. 2020) and D.T2.2.5 (Polin et al. 2020).

[4] Cf. *trAILs* Deliverable D.T2.2.6 (Bär et al. 2020).

[5] Cf. *trAILs* Deliverable D.T2.2.2 (Krošelj et al. 2019).

6.3 Test-Design Workshop

Preliminary activities

The responsible project partner (CAUE 84), in close collaboration with CAUE 05 and the *Communauté de communes du Pays des Ecrins*, set up the workshop organization in advance, including a detailed meeting schedule (three days, of which the second one with stakeholders), inputs and expected outputs. The agenda was shared with the partners a couple of weeks before the event.

As design-based input material, useful to trigger the discussion around the roundtables, TUM provided three design proposals for the site, all advancing a rather radical transformation/redevelopment. Furthermore, the site assessment booklet containing all the thematic analyses was prepared by UL and printed in view of the workshop activities.

In terms of set-up, the workshop format was built around three interdisciplinary tables (i.e. working groups) focusing each on one specific design proposal, while addressing a common list of issues or elements for discussion: current and future activities on the site; linking the site/district to the Durance river; relationships with the town center and the railway station; enhancement of the industrial heritage on the site; relationship between urban and natural mountain landscape; use of existing buildings and structure; pollution and natural risk management; targeting of the area as a priority development site for the region; renewable energies production.

A detailed list of stakeholders to be involved was previously drafted and around 40 among stakeholders and observers were contacted in advance, aiming at involving as more target groups as possible: municipality (mayor, local council, head of administrative/planning sectors), representatives of other neighboring municipalities, representatives of major cultural, tourist and environmental associations, regional development agencies, members of the civil society, representatives of institutions/fields related to brownfield rehabilitation. Working materials to be arranged on the tables were provided in advance.

Workshop settings

The workshop was planned as a direct follow-up to the previous brownfield co-assessment meeting in L'Argentière-la-Bessée and La-Roche-de-Rame (November 2019), during which the project partners, local elected officials and local/regional experts visited both sites and briefly discussed the potential for redevelopment. Together with the regional representatives, it was decided to focus, during the workshop, on the first site only in L'Argentière-la-Bessée, as the most demanding and challenging among the two.

Before starting the roundtable discussion, a short video produced by UNIVR[6] was played to be seen by all the participants. The video described the everyday life in L'Argentière-la-Bessée and La-Roche-de-Rame, focusing on issues such local identity, the impact of deindustrialization, the current living conditions and the prospects

[6] In cooperation with INTWIG communication and media agency.

or expectations for the future. The video helped the stakeholder to get engaged in the workshop activities. Furthermore, the design proposals were briefly presented by TUM to the whole group too, together with a poster summarizing the results of the assessment reports assembled by the project partners with the coordination of UL (policy analysis, territorial and spatial analysis, socioeconomic analysis, ecological assessment).

After these opening presentations, the interdisciplinary working groups were formed by matching research partners with the 21 participating stakeholders[7] and moderators/translators, being then invited to their respective tables to start the discussion.

The exchange within the working groups was sustained by an open framework of questions, focusing on key aspects such as the perceptions of the site in relation to the context (town and valley) or the issues raised by the design proposal on the table, or impacts of the site redevelopment program. The distribution of local and regional stakeholders around the tables aimed at mixing profiles in order to encourage the confrontation among different approaches to the site and its future. Each table was provided with the same working materials: a poster summarizing the results of the thematic assessments, aerial photos, a map of public and private properties, a poster describing the design proposals (one per table). The results from the working tables were included into a sketched 'programmatic plan' (Fig.6.5), generated from a general and summarizing discussion among the table participants.

In the second part of the second day, finally, the three programmatic plans produced by the working groups were revised and discussed in a plenary session. This fundamental step allowed to compare the work of the tables while synthesizing the different orientations, thus identifying common key issues for the site transformation.

Roundtable 1

Group: 8 stakeholders, 7 research partners, 1 interpreter, 1 moderator.

Enhance the heritage value of the former hydroelectric power plant.

[7] The participating stakeholders were: Cyrille Drujon D'astros (President of CC Pays des Ecrins—Mayor of Fréssinière), Patrick Vigne (Mayor of l'Argentière), Gérard Guimbert (Deputy mayor of l'Argentière), Marie-Pierre Bocchiardo (elected official—municipality of l'Argentière), Stéphanie Davin-Poncelet (CC Pays des Ecrins—Economic development), Philippe Collet (CC Pays des Ecrins—Economic development), Julie Privat (CC Pays des Ecrins—Tourism department), Bettina Matias (CC Pays des Ecrins), Raymond Marigne (Vice-chairman CAUE 05, Architect), Xavier Grojeanne (Public Land Institution PACA), Valérie Cécile Chauvin (SUD PACA Region—Directorate for Energy Transition and Territories/observer), Cédric Reynaud (SUD PACA Region—Economic development), Florent Ferrucci (AREA—SUD PACA Region), Olivier Chazaud (Chambre de métiers PACA—Hautes Alpes Delegation), Frédéric Sabatier (Urbanist—Parc National des Ecrins, planning department), Regis Allamanno (Building and public works contractor, Sablonnières business park), Daniel Girard (citizen of l'Argentière—former employee AFP site), Guido Baschenis (Piedmont Region/observer), Elsa Giraud (L'Atelier d'Histoire—Historian, cultural engineer and tour guide), Prune Saez Contreras (Architecture student at ENSA Grenoble), Gaëlle Privat (Architecture student at ENSA Lyon).

- Preserve the power plant building as the identity of the site and the city (heritage) (priority).

Developing new activities on the site, for a mix of urban uses.

- Analyze the current uses of the site to assess its place in the context of the city and estimate the needs.
- Build on the current situation and uses to define a new identity for the site.
- Establish a cultural center on the façade of the main artery (on the site of the AFP brownfield), linking the town center/station and the Sablonnières district.

Restructuring and enhancing the neighboring built environment.

- Preserve/enhance or erase existing buildings regarding their heritage value and for the identity of the site and the city (priority).
- Build on the current situation and uses to define a new identity for the site.
- Establish a cultural center on the façade of the main artery (on the site of the AFP brownfield), linking the town center/station and the Sablonnières district.

Link the site to the Durance River/Study the possibilities of transforming the «Serie E» building in this direction.

- Highlight in the development program the water/Durance element as the identity of the place/city (local community/economy), integrating well the flood risks.

Strengthen soft modes urban links in and around the site by creating a network of structuring public spaces.

- Improve the soft mode accessibility of the district and the connections with the different points of interest in the city/site (including PMR)—Create a central axis of soft mobility, linking the town center to the leisure areas in the south and structuring the district (priority).

Integrate and link the Argentière—les Ecrins station sector to the site requalification project. [no suggestions].

Know and manage site pollution.

- Clarify the pollution situation at the site and implement the necessary clean-up actions (priority).

Reinforce the park/natural area character of the southern part of the site.

- Make the most of the natural area to the south, linking it to the city by a structured cyclo-pedestrian axis running through the Sablonnières district.

Roundtable 2

Group: 5 stakeholders, 7 research partners, 1 interpreter, 1 moderator.

Enhance the heritage value of the former hydroelectric power plant.

- Begin the project to requalify the area with the former hydroelectric power plant and by enhancing the heritage buildings (priority)—Demolish the buildings around the power plant to enhance it.

Developing new activities on the site, for a mix of urban uses.

- Giving a mix of functions to the neighborhood (economic activities, shops, culture, housing, etc.) (priority).
- Focus the development of the CCPE's urbanization on this site, densify it (priority).
- Develop activities contributing to sustainable development (short circuits, renewable energies, etc.).

Restructuring and enhancing the neighboring built environment.

- Focus the development of the CCPE's urbanization on this site, densify it (priority).

Link the site to the Durance River/Study the possibilities of transforming the «Serie E» building in this direction.

- Restore the banks of the Durance to the inhabitants and to reduce the risk of flooding (priority).

Strengthen soft modes urban links in and around the site by creating a network of structuring public spaces.

- Create a good connection with the town center—Plan a global requalification project that integrates the sub- division area north of Les Sablonnières and the station area.

Integrate and link the Argentière—les Ecrins station sector to the site requalification project.

- Strengthen links with the station area.

Know and manage site pollution.

- Remediate soil pollution and enhance the natural environment of the site.

Reinforce the park/natural area character of the southern part of the site.

- Affirm the character and create a dialogue between the 2 parts of the site: urban (in the north) and natural (in the south) and the district as a whole with the wider landscape.
- Bring the park space closer to/extend it to the contact with the built front of the business district, creating a clear break in urbanization on this boundary.

Roundtable 3

Group: 5 stakeholders, 8 research partners, 1 interpreter, 2 moderators.

Enhance the heritage value of the former hydroelectric power plant.

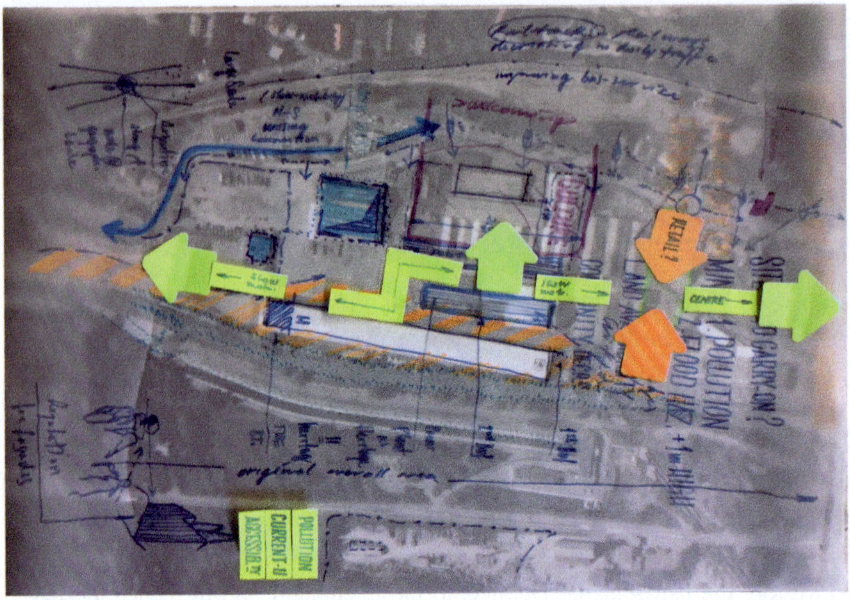

Fig. 6.5 The preliminary (draft) key map resulting from the discussion at Table 1. © Massimo Galluzzi

- Requalify the building of the power plant, clear the surroundings to enhance it and connect it to the neighborhood (priority).

Developing new activities on the site, for a mix of urban uses.

- Bring a mix of uses to the site, in particular by setting up a center for the socio-cultural activities of the town/ community of communes (priority),
- Develop economic activities related to the urban setting of the site—Structure this activities development on the CCPE in connection with the proposed economic activity zone of La Roche-de-Rame (outside the conurbation).
- Define a method and a strategy to build the project (priority),
- Implement innovative activities in the continuity of the culture of innovation specific to the site and to this mountain region.

Restructuring and enhancing the neighboring built environment.

- Organize the requalification of the district, its connections with the surrounding elements, around one or two structuring public spaces,
- To develop innovative architectural programs in line with the culture of innovation specific to the site and this mountain region.
- Define a method and a strategy to build this project (priority).

Link the site to the Durance River/Study the possibilities of transforming the «Serie E» building in this direction.

- Remove or transform the «E-Series» building to enhance the site by opening it to the river and create a buffer zone to better manage floods (priority).

Strengthen soft modes urban links in and around the site by creating a network of structuring public spaces.

- Strengthen the link with the town centre, the station and nature areas around an axis of structuring public spaces (priority).

Integrate and link the Argentière—les Ecrins station sector to the site requalification project.

- Connecting the station site to the Sablonnières district by structuring developments (priority).

Know and manage site pollution.

- Know the pollution of the site prior to the project.

Reinforce the park/natural area character of the southern part of the site.

- Reinforce the city/nature gradient from north to south,
- Establish a public space between the built-up area and the green zone?

6.4 Planning Recommendations

In the case of L'Argentière-la-Bessée, the planning recommendations emerged already during the workshop, thanks to a very detailed draft key map resulting from the plenary conclusive discussion. The comparison between the three different outcomes from the working tables led, in fact, to outline a comprehensive and clear planning scheme based on mutually agreed issues for the site transformation. In the post-workshop phase, the efforts were concentrated into 'turning back' the draft key map into different development paths, according to the intensity of transformation discussed during the workshop: a soft transformation (start-up/temporary scenario) as well as two radical ones ('back-to-city' and 'back-to-nature'), besides the revised key map. For each development path, the key actions related to the structuring issues for the site transformation—ecological network and public open spaces (including potentially relating built-up ones)—are described with text and an attached illustrative plan.

Start-up/temporary activation

In view of a comprehensive transformation of the former industrial area, the reactivation of some of its parts might help, as a priority intervention, to trigger the whole regeneration process. The 'start-up' path envisages specific actions aimed at preparing the area, such as the preservation, refurbishment and adaptive reuse of the heritage buildings (former hydroelectric power station, ancient atelier Pechiney,

Table 6.4 Start-up scenario, key issues

Ecological network	*Landscape-environmental recovery of Durance riverbanks*: promote facilities on the banks, pathways, landscaping, etc.
	Durance river: manage the flooding risk on the site by creating a buffer zone between the river and the inner portions of the site
	Green connections: improve the green network by linking built-up areas, the river and leisure areas in the south
	Blue connections: improve physical, functional and visual connections between the inner site and the river landscape
Public open spaces	*Public open spaces and pathway network*: connect the activated parts of the site and enhance the livability of the area
	Industrial heritage recovery and enhancement (hydroelectric power station, atelier Pechiney, penstocks): preserve the historical built structures as site landmarks and prepare to host new functions

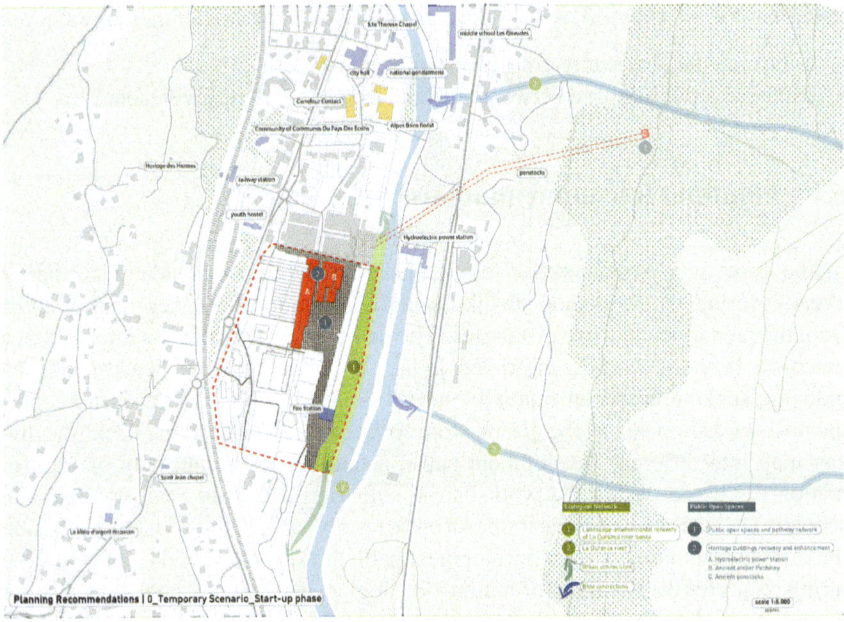

Fig. 6.6 Start-up scenario. © Politecnico di Milano/Interreg project *trAILs*

water pipelines/penstocks) as well as measures to open up the area to river landscape. This involves the identification and development of a minimal ecological and mobility/open spaces network, in order to connect the activated parts of the site and the outer areas (Table 6.4; Fig. 6.6).

Intensive development/ *back to city*

The 'back-to-city' development path addresses a substantial redevelopment of the site through the reuse and reactivation of preserved industrial buildings as well as by defining new building areas and adding new volumes and built structures. To this purpose, the site internal layout is refurbished by setting-up of a network of new paths, public spaces and servitudes that integrates coherently the existing structure, while ameliorating the ecological conditions of the whole area (Table 6.5; Fig. 6.7).

Extensive development/'back to nature'

The 'back-to-nature' development path deals with the foreseeable perspective of a progressive underusage of the site. The conservation of just a few relevant built structures (identified as industrial heritage) coexists with an extensive environmental

Table 6.5 Intensive development scenario, key issues

Ecological network	*Landscape-environmental recovery of Durance riverbanks*: promote facilities on the banks, pathways, landscaping, etc.
	Durance river: manage the flooding risk on the site by creating a buffer zone between the river and the inner portions of the site, as well as to protect the adjacent urban neighborhoods
	Increasing of vegetation cover: enhance the semi-natural look of the site by expanding and improving the actual vegetation cover, in line with the built/green ratio of the surroundings
	Improve succession areas: ecological restoration on former slag heap and surrounding leftover areas in the south, including the improvement of natural amenities as well as a cycle/pedestrian path along the river to connect the leisure center nearby
	Green connections: connect green areas/river with built areas, as well as leisure areas in the south
	Blue connections: improve physical, functional and visual connections between the inner site and the river landscape
	Green roofs: improve the thermal and acoustic efficiency of the buildings and help to counteract the effects of waterproofing urban settlements
Public open spaces	*Public open spaces and pathway network*: connect the activated parts of the site and enhance the livability of the area
	Industrial heritage recovery and enhancement (hydroelectric power station, atelier Pechiney, penstocks): preserve the historical built structures as site landmarks and prepare to host new functions
	Adaptive reuse of existing buildings: foster the adaptive reuse of old industrial buildings by implementing innovative architectural programs, in order to restore the site attractiveness as a place of cultural and technological innovation
	New buildings: to host new functions for the site future development, new buildings should be realized around the historical core of preserved heritage structures

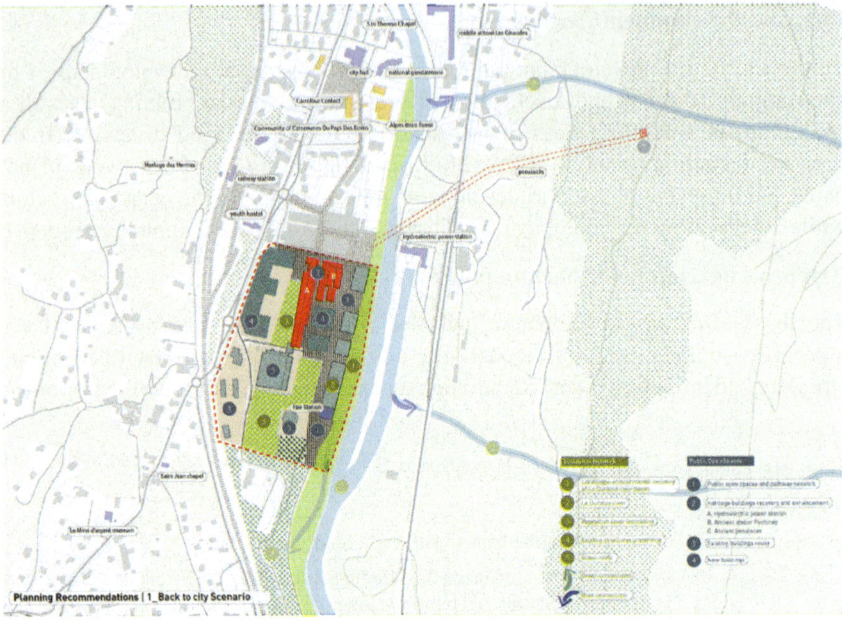

Fig. 6.7 Intensive development scenario. © Politecnico di Milano/Interreg project *trAILs*

regeneration of the former industrial area, in coherence with the surrounding landscape. The renaturation of the site involves three main phases and relative regenerative landscapes: (1) sparsely vegetated landscape with spontaneous succession, (2) grassland-dominated landscape, (3) forest-dominated landscape (Table 6.6; Fig. 6.8).

Key Map

As shared between the three development paths, and preliminary to their future implementation, the key map integrates and illustrates the commonly agreed issues as identified in the test-design workshop. These include the essential network of public spaces, the environmental system and relevant ecological spaces as well as the most relevant buildings and structures to be preserved and reused. By displaying the basic structure for the future development of the site, the key map is a valuable tool in the hands of the community, a starting point for any possible transformation. The minimum requirements for the site transformation and regeneration, no matter what will be the scenario that the community implement, are listed here below (Table 6.7; Fig. 6.9).

Table 6.6 Extensive development scenario, key issues

Ecological network	*Landscape-environmental recovery of Durance riverbanks*: promote facilities on the banks, pathways, landscaping, etc.
	Durance river: manage the flooding risk on the site by creating a buffer zone between the river and the inner portions of the site, as well as to protect the adjacent urban neighborhoods
	Increasing of vegetation cover: enhance the semi-natural look of the site by expanding and improving the actual vegetation cover, in line with the built/green ratio of the surroundings
	Improve succession areas: ecological restoration on former slag heap and surrounding leftover areas in the south, including the improvement of natural amenities as well as a cycle/pedestrian path along the river to connect the leisure center nearby
	Green connections: connect green areas/river with built areas, as well as leisure areas in the south
	Blue connections: improve physical, functional and visual connections between the inner site and the river landscape
Public open spaces	*Public open spaces and pathway network*: connect the activated parts of the site and enhance the livability of the area
	Industrial heritage recovery and enhancement (hydroelectric power station, atelier Pechiney, penstocks): preserve the historical built structures as site landmarks and prepare to host new functions

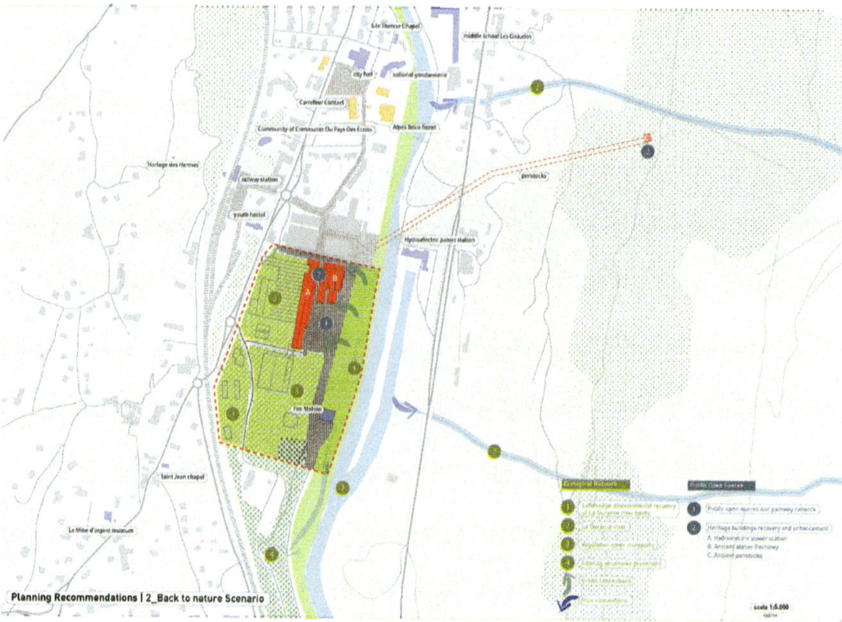

Fig. 6.8 Extensive development scenario. © Politecnico di Milano/Interreg project *trAILs*

Table 6.7 Key map issues

Landscape-environmental recovery of Durance riverbanks: promote facilities on the banks, pathways, landscaping, etc.
Durance river: manage the flooding risk on the site by creating a buffer zone between the river and the inner portions of the site, as well as to protect the adjacent urban neighborhoods
Increasing of vegetation cover: enhance the semi-natural look of the site by expanding and improving the actual vegetation cover, in line with the built/green ratio of the surroundings
Improve succession areas: ecological restoration on former slag heap and surrounding leftover areas in the south, including the improvement of natural amenities as well as a cycle/pedestrian path along the river to connect the leisure center nearby
Green connections: connect green areas/river with built areas, as well as leisure areas in the south
Blue connections: improve physical, functional and visual connections between the inner site and the river landscape
Public open spaces and pathway network: connect the activated parts of the site and enhance the livability of the area
Industrial heritage recovery and enhancement (hydroelectric power station, atelier Pechiney, penstocks): preserve the historical built structures as site landmarks and prepare to host new functions
New building zones: identify appropriate building zones for the site future development, to be concentrated around the historical core of preserved heritage structures

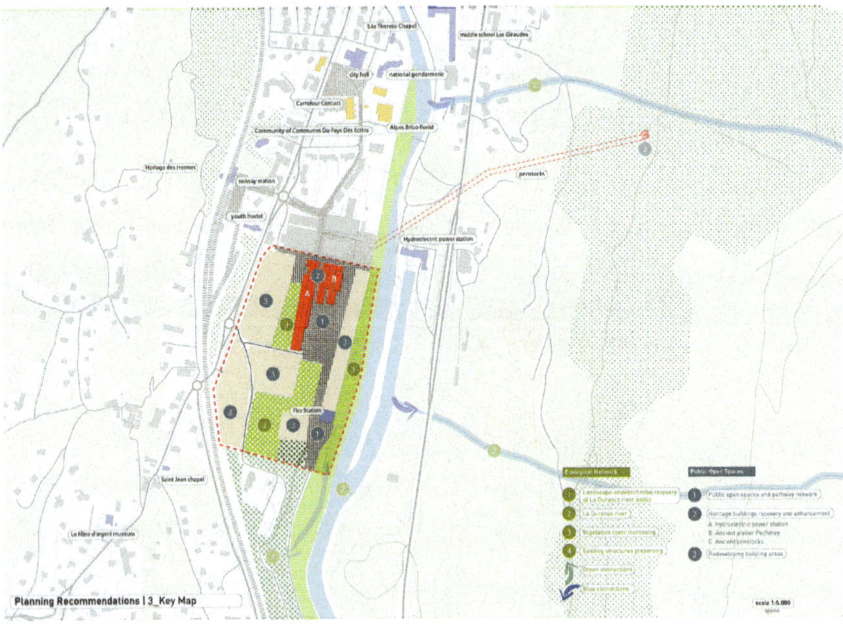

Fig. 6.9 Key map. © Politecnico di Milano/Interreg project *trAILs*

References

Bär K, Kollmann J, Kleitz T (2020) EU Interreg project trAILs—Deliverable D.T2.2.6.—Environmental assessment report L'Argentière-la-Bessée. Available at: https://www.alpine-space.org/projects/trails/pilots/_results/wp-t2/d_t2_2_6_environmental_assessment-report_france.pdf

Kleitz T (2018) EU Interreg project trAILs—Deliverable D.T1.3.2.—Pilot Profile L'Argentière-La Bessée. Available at: https://www.alpine-space.org/projects/trails/pilots/_results/wp-t1/pilot_profile_fr.pdf

Krošelj M, Pipan T, Marot N et al (2019) EU Interreg project trAILs—Deliverable D.T2.2.2.—Policy assessment report L'Argentière-la-Bessée. Available at: https://www.alpine-space.org/projects/trails/pilots/_results/wp-t2/d_t2_2_2_policy-assessment_regional_report_france.pdf

Polin V, Prete V, Kleitz T (2020) EU Interreg project trAILs—Deliverable D.T2.2.5.—Economic assessment report L'Argentière-la-*Bessée.* Available at: https://www.alpine-space.org/projects/trails/pilots/_results/wp-t2/d_t2_2_5_economic_assessment-report_final.pdf

Solero E, Vitillo P, Galuzzi P et al. (2020) EU Interreg project trAILs—Deliverable D.T2.2.3.—Spatial planning and landscape assessment report L'Argentière-la-Bessée. Available at: https://www.alpine-space.org/projects/trails/pilots/_results/wp-t2/trails_d_t2_2_3_spatial_assessment_france_final.pdf

Veronesi L, Migliorati L, Lanfranchi G et al (2020) EU Interreg project trAILs—Deliverable D.T2.2.4.—Socio-demographic assessment report L'Argentière-la-Bessée. Available at: https://www.alpine-space.org/projects/trails/pilots/_results/wp-t2/trails_d_t2_2_4_social_assessment-france_final_univr-caue_final.pdf

Chapter 7
Tržič, Slovenia

The chapter presents the Slovenian pilot site through the activities and achieved results in the framework of the trAILs project. A brief overview of both the regional context and the brownfield site is first provided, describing the main features and transformation challenges. Secondly, the identified potentials and problems are listed, as emerged from to the spatial assessment and the site visit. The chapter describes then the test-design phase, focusing on the workshop setting and methodological aspects and procedures. Finally, the planning recommendations for the site are presented, as outlined during and after the workshop.

7.1 Regional and Site Overview[1]

The local community of Tržič lies in the north of the Gorenjska cultural and statistical region in Slovenia, bordering Austria. The community covers 155.4 km^2 (0.8% of the territory of the Republic of Slovenia) and includes 35 settlements with a total of 15,851 inhabitants (2007). The center of the local community is the historic town of Tržič, surrounded by the Karawanks mountain range to the north and the prealpine forelands of the Sava valley to the south. The region has a typical Alpine climate and land use, with 70% of the territory covered by forests and farmland. The mountains and hills in particular face avalanche risk in the winter, while flooding and landslides often occur in the valley floor during times of heavy rainfall. The Tržič region has good transport connections via roads toward Austria and the capital city, Ljubljana. The national airport Jože Pučnik (Brnik) is roughly half an hour away by car. The inhabitants are well educated, but education facilities, except primary school, must in general be accessed in other cities, in the region and in the capital city Ljubljana. Economically speaking, the region and especially the town of Tržič have been a key center of craft and industry in the previous centuries and still are today.

[1] Information retrieved and adapted from *trAILs* Deliverable D.T1.3.2 (Cvenkel et al. 2019).

M. Modica and E. Solero, *Brownfield Transformation in Fragile Territories*,
PoliMI SpringerBriefs, https://doi.org/10.1007/978-3-031-09083-7_7

Larger, traditional companies—factories producing shoes, metal objects, textiles, paper, furniture—were until 1990 an important generator of development and income in the local community. After Slovenia became independent in 1990, several companies lost their competitive placement in the new market, ending up in bankruptcy after years of changes in management and production. Many former employees who were employed in these traditional factories started their own businesses, many of which are now developing successfully while profiting from the local manufacturing know-how and expertise. The town of Tržič is surrounded by nature and the region has several protected areas (Natura 2000), especially in the Karawanks mountain range and Dolžanova gorge. For this reason, tourism has become in recent years an emerging and promising economic sector.

The site of Bombažna Predilnica in Tkalnica (BPT, 'cotton spinning and weaving mill' in English), one of the most significant remnants of the local industrial past, lies close to the old town of Tržič (Fig.7.1). The textile company BPT was established by the Swiss merchant Edmund Glanzmann and his Austrian partner Andrew Gassner at the end of nineteenth century. It began production back in 1886. Besides the production of textiles, the factory produced electricity through its own four hydroelectric power stations. In 1922–23 the factory employed 1100 people, having already 72 houses of property to host them. At the end of twentieth century, due to loss of market and obsolescence, BPT ceased to exist. The various limitations and constraints of the disused space meant that it took almost 20 years for the new owning company

Fig. 7.1 The BPT site holds several industrial heritage buildings and structures. © Občina Tržič

to plan the revitalization of the area, in effective cooperation with local and regional stakeholders. The local community of Tržič owns today a small part of the area, on which it is developing services that will contribute to a better quality of life in the town, for example by creating green areas. The site itself is a combination of historic, protected or not, heritage buildings, a power station, a network of internal roads and small green leftover spaces. The site is connected to the main road, that is in turn linked to the center of Tržič and leads to villages in the surroundings. The site is indeed well connected to the road network leading to Austria as well as towards the Slovenian national airport and the city of Ljubljana. Communal services and IT connections are also available on part of the site.

For a successful revitalization of the BPT site it is essential to emphasize and reinforce the infrastructural and cultural connections to the hosting center of Tržič. Since BPT is closely connected with the town's historic core, it shares certain supply sources, development activities and activities connected to quality of life and entrepreneurship. The site has a lot of potential and requires limited efforts to be properly reused, being a mixture of old yet valuable buildings and structures which need only general improvement (no demolition), plus open spaces suitable to be redesigned. The site represents therefore a unique opportunity for redeveloping the entire city of Tržič as well as for relaunching the local and regional identity in a sustainable way.

7.2 Spatial Assessment[2]

The outbreak of the Covid-19 pandemic severely affected both on- and off-site activities related to the Slovenian pilot site, since most of the planned meetings, workshops and fieldwork had to be adapted to online formats.

The assessment of the spatial conditions of the site and its context began with a 'virtual' site visit, which took place on 18–19 June 2020. During the first day, organized by BSC Kranj and UL, the project partners were virtually guided by local representatives and experts on the BPT site through a video tour of the area and the town of Tržič. Then, UL presented a brief overview of the design challenges and activities through the experiences of its own students. Research partners were then invited in advance to write down questions to be submitted to the stakeholders during the reflection session scheduled for the following day. In the morning, the project consortium had a chance to virtually meet the local stakeholders (municipality officials, BPT site owner, BPT site users) in a round table session. The main goals consisted in understanding which assessment the stakeholders considered as useful for the sites and how these assessments should be framed, in terms of problems and

[2] Cf. *trAILs* Deliverable D.T2.2.3 (Polytechnic University of Milan and Vienna University of Technology 2020).

potentials of the site. This exploratory discussion was extremely useful to the partners to draft their thematic assessments and the assessment tool, the final methodological deliverable.

Concerning the spatial assessment itself, the knowledge gained during the visit and the roundtable with the stakeholders was then integrated, off-site, with the local planning documents provided by the regional partners. Different aspects were covered, from the regional accessibility to the local conditions of the site and its context. The resulting spatial assessment is presented below through its three main analytical frameworks: environment, settlements and mobility. Further analyses were also done (e.g. planning rules and constraints, supply and disposal infrastructures—electricity, gas, water supply networks) and used in the workshop.

Environment

The brownfield and its immediate surroundings are mostly characterized by buildings and paved, impermeable open spaces. The site is enclosed by canalized water courses (river and canals) and does not include any area with significant environmental qualities. The context presents a diverse environmental pattern, with the highest ecological value registered on the forested mountain slopes than in the valley floor adjacent to the site. The situation is portrayed in a specific map (Fig. 7.2) and described through its potentials and problems in Table 7.1.

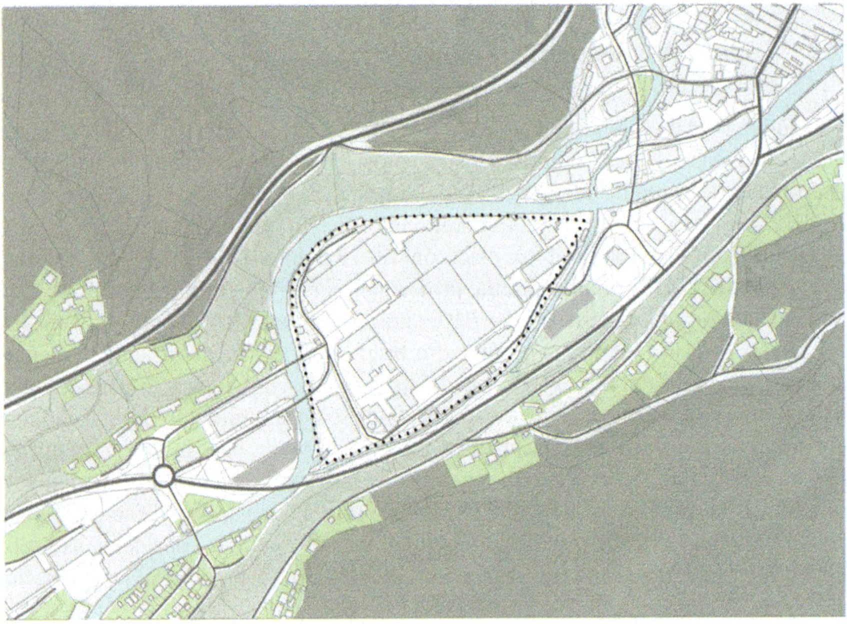

Fig. 7.2 Spatial assessment, environment map. © Politecnico di Milano/Interreg project *trAILs*

Table 7.1 Spatial assessment, environment-related potential and problems

Potentials	Problems
Diverse landscape, with high potential in terms of biodiversity and ecological connections as well as for touristic and recreational purposes	Climate change affecting the water regime of the Tržič Bistrica river, thus increased risk of natural hazards such as flooding for the site and the town center too
Many water bodies (rivers, streams, canals) nearby the site and in the region (Sava is the largest river of both Slovenia and Croatia)	Unfavorable topography of the site
Natural protected areas in the vicinity (Natura 2000—habitats directive areas and Natura 2000—birds directive areas)	
Location of the site along the Tržič Bistrica river, thus high potential for environmental valorization of the site in connection to the river corridor (valley-wide green infrastructure)	

Settlement structure

The site is characterized by old industrial buildings with a very large build footprint compared to those in the surrounding. The brownfield site is very well embedded in the urban system of Tržič and its valley, which implies a proximity to the historic town centre (less than 500 m of distance). This aspect can be seen as a chance for renovating both the site and the whole old town, in a joint transformation process. The

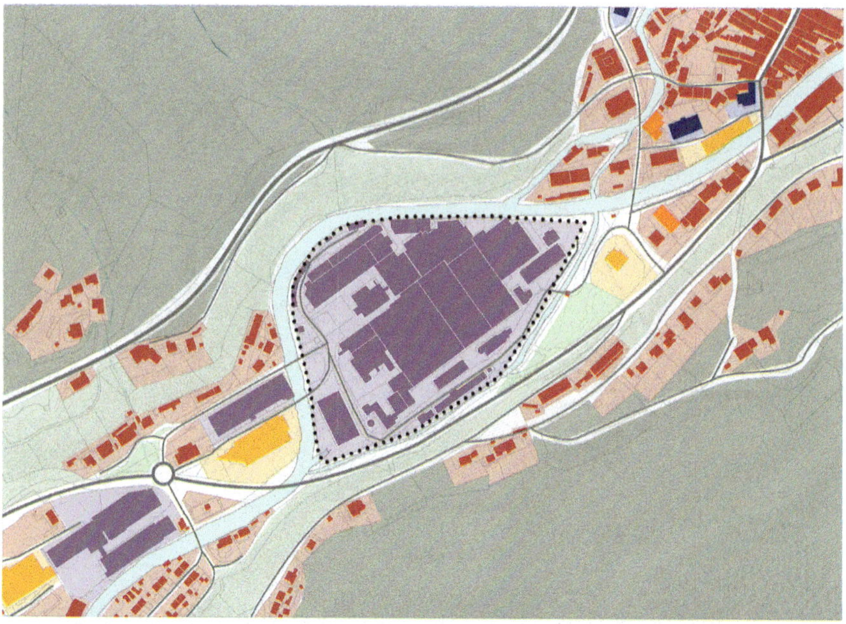

Fig. 7.3 Spatial assessment, settlements map. © Politecnico di Milano/Interreg project *trAILs*

Table 7.2 Spatial assessment, settlements-related potential and problems

Potentials	Problems
The site is centrally located within Tržič, within walking distance from main services and facilities	Excessive interventions in the stock of industrial buildings may weaken the identity of the town as a large industrial site
Good level of integration of the site within the existing urban structure	Agricultural areas are occupied by mainly commercial zones, while large areas, previously occupied by industry, are now empty or disused
The site has a compact urban footprint, favorable to inner development and densification	The site is enclosed by water courses, posing the risk of creating an isolated island in the middle of the city center

settlement structure of the surroundings is characterized by scattered residential areas (sprawled hamlets) on the slopes and compact urbanization in the valley floor. The situation is portrayed in a specific map (Fig. 7.3) and described through its potentials and problems in Table 7.2.

Mobility

Due to its central location, the site is well connected to the road system and the local network of public transport. The nearest railway stations, however, are in Kranj and Radovljica, from which Tržič can be reached in half an hour by bus. The situation is portrayed in a specific map (Fig. 7.4) and described through its potentials and problems in Table 7.3.

Fig. 7.4 Spatial assessment, mobility map. © Politecnico di Milano/Interreg project *trAILs*

Table 7.3 Spatial assessment, mobility-related potential and problems

Potentials	Problems
High accessibility by road and bus lines, including seasonal public mobility for touristic purposes	No railway service for passengers
Good location as a local 'gateway' from Central to Southern Europe	No railway infrastructure for freight transport

In the view of the workshop, the spatial assessment was then integrated by the other thematic assessments carried on by the scientific partners, according to their own specialized expertise. The results are here briefly mentioned with reference to the thematic focus.

Socio-economic analysis[3] *(UNIVR* the local community is quite cohesive, though its population is registering since decades a decreasing trend. A participatory approach to the BPT site transformation is required and highly welcomed, to keep local community alive as well as to provide new impetus to the local economic dynamics.

Ecological analysis[4] *(TUM)*: despite lacking significant natural ecologies, due to the mostly built-up surfaces, the site is a local hotspot for several flora and fauna species. The conservation and improvement of habitats is recommended, in particular some rare and legally protected animal species identified inside the area and close to existing water bodies. No contamination is reported.

Policy analysis[5] *(UL)*: at the regional and local level there is a lack of understanding of brownfields as an opportunity for sustainable urban and territorial regeneration. A long-term planning and financial support for brownfield redevelopment is missing. At the same time, a local planning vision for the BPT area is well-established, including detailed approved plans and documents.

7.3 Test-Design Workshop

Preliminary activities

In the context of the pandemic, and due to the related restrictions, the project team decided to opt for a hybrid format for the test-design workshop, with only the stakeholders and local partners on the site while the rest connected online. The workshop agenda was set up in advance by UL in close collaboration with BSC Kranj, including a detailed meeting schedule (two days, of which the first one with stakeholders), inputs and expected outputs. Before drafting the agenda, the Slovenian partners organized a virtual meeting with POLIMI (WP T3 lead) in order to share

---

[3] Cf. *trAILs* Deliverables D.T2.2.4 (Lanfranchi and Migliorati 2020) and D.T2.2.5 (Polin and Prete 2020).

[4] Cf. *trAILs* Deliverable D.T2.2.6 (Bär and Kollmann 2020).

[5] Cf. *trAILs* Deliverable D.T2.2.2 (Krošelj et al. 2019).

the planned organization and settings. The agenda was shared with the partners a month before the event.

As design-based input material, useful to trigger the discussion around the roundtable, TUM provided five design proposals derived from its own teaching studio with landscape architecture students. Through a comparative matrix, designed by POLIMI, the three most significant proposals (different options from a soft to a radical site transformation) were selected to be used as inputs in the workshop (Fig.7.5). In terms of organization, the agreed format was oriented towards having one single interdisciplinary table focusing in turn of each of the proposals.

A detailed list of stakeholders to be involved was previously drafted and around 20 among stakeholders and observers were contacted in advance, aiming at involving as more target groups as possible: municipality (mayor, local council, head of administrative/planning sectors), representatives of major cultural, tourist, environmentalist association, members of institution related to the cultural heritage and economic sector. Working materials to be arranged on the tables were provided in advance.

Workshop settings

The platform chosen for the workshop was GoToMeeting™. The first day was divided into three sessions. In the first one, after the welcome greetings and presentations, the research partners shortly presented the results of the assessment (policy analysis, territorial and spatial analysis, socioeconomic analysis, ecological assessment) while TUM presented the three selected design proposals. Before the start of the roundtable discussion, a short video produced by UNIVR[6] was played to be seen by all the participants. The video described the everyday life in Tržič, focusing on issues such local identity, the impact of deindustrialization, the current living conditions and the prospects or expectations for the future. The video helped the stakeholder to get engaged in the workshop activities.

In the second session, the roundtable discussion with the participating stakeholders (regional partners and 10 stakeholders[7]) took place exclusively in Slovenian language. The three proposals were presented and discussed according to a couple of key questions: (1) what are the opportunities and challenges of the proposal for the BPT area? (2) what happens to the city center if the proposed transformation is realized? The working materials on the table consisted in a book summarizing the results of the thematic assessments, aerial photos and the three selected proposals.

[6] In cooperation with INTWIG communication and media agency.

[7] The participating stakeholders were: Bojan Knific (Museum of Tržič), Daniela Žagar (The Chamber of Craft and Entrepreneurship of Slovenia), Drago Perc (Employment Chamber of Slovenia), Mojca Brejc (Slovenia Forest Service), Željko Babič (municipality of Tržič, department of road and infrastructure) Klemen Srna (municipality of Tržič, department of economic development), Katja Rekelj (Municipality of Tržič, department of economic developement), NataŠa Ulen (Public Institute of the Republic of Slovenia for the Protection of Cultural Heritage), Jon GrobovŠek (Public Institute of the Republic of Slovenia for the Protection of Cultural Heritage), Metod Rogelj (Institute of the Republic of Slovenia for Nature Conservation).

In third session, the Slovenian partners presented the results of the workshop to the project partners—which were connected online. A short exchange between the stakeholders and the project team then followed.

Despite the difficulties related to a hybrid workshop format, the overall workshop proved to be successful: many opinions, reflections and constructive proposals were expressed and debated during the meeting.

Roundtable

Group: 10 stakeholders (on site), 1 regional partner (on site), 1 coordinating research partner (online), 8 project partners (online).

Scenario description (1-first scenario, 2-s scenario, 3-third scenario, C-cross features).

- (1) Gradual site recovery through a mix of micro and small enterprises, cultural activities, tourism and services.
- (2) Monofunctional site-wise transformation, through large investments on new productive activities.
- (3) Site deconstruction through removal of most of the buildings, preserving a few symbolic ones for touristic purposes and leaving the rest as awaiting green/renaturalized space with no additional functions.

Nature and green surfaces

- (1) Smaller companies tends to appreciate and care more the green spaces in the site and around.
- (2) A quality design of open and green spaces represents a value for industry/productive activities.
- (3) Opening up the area is a positive fact, especially in relation to the possibility of micro reforestation, recreation activities and general attractiveness of the site for visitors.
- (C) Green spaces on the site are important and must be valued/increased.

Economy

- (1) Very realistic development option for the site, including gradual and small-scale transformations and focusing on crafts and technology.
- (2) Potentially new added-value jobs, but also unrealistic that the site will be occupied by just one mayor player/activity.
- (3) Urban green areas are welcome and might increase the attractivity of the site/town/region.
- (C) -

Cultural heritage

- (1) Small-scale transformations are positively impacting on cultural heritage, since they don't change the site appearance and structure. The project could help to strengthen local branding/marketing strategies.

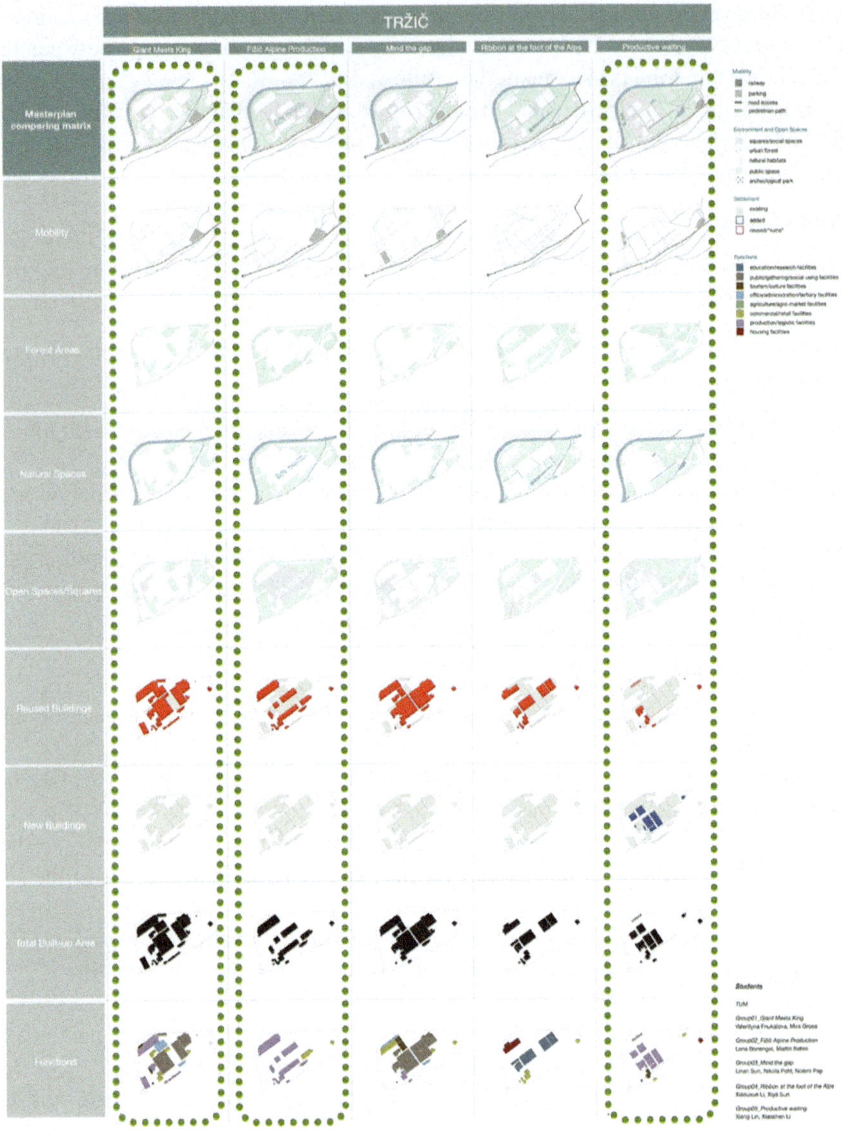

Fig. 7.5 The comparative matrix for selecting the design proposals (highlighted in green the three chosen ones)

- (2) One key player for the whole site is likely to develop massive interventions, which are not desirable in terms of cultural heritage. In addition, if one big player fails the redevelopment, the site is compromised. The 'closure' of the site to the outside is not desirable as could prevent cultural heritage to be appreciated/seen.
- (3) -
- (C) The cultural heritage of the site is quite undervalued.

Local community, sense of place

- (1) (2) Small-scale activities and transformations are helpful to bring the community closer to the site redevelopment.
- (3) Empty green spaces are not really helpful to keep the local community economically active.
- (C) -

Quality of living, environment

- (1) (2) Beware of emphasizing activities that require a lot of space and efforts to be implemented, as the local quality of living might be affected.
- (3) Interventions beyond the scope of the local development plan and the site conservation plan are not acceptable.
- (C) The connection between the site and the town must be strengthened, visually and conceptually.

Employment

- (1) Reusing existing infrastructures and spaces is a plus, as it allows new employment to start quickly.
- (2) The proposal is positively valued, as new jobs are welcome. The site could absorb around 200 local inhabitants, currently unemployed.
- (3) -
- (C) -

Education and young generations

- (1) -
- (2) From the point of view of younger generations who want to be employed locally, it makes more sense to have smaller companies with diversified activities on the site rather than just one big employer.
- (3) High quality urban green on the site is welcome to attract young people.
- (C) Missing links between the site potential offer for young people and the existing local situation.

Tourism potential

- (1) The site might be opened for tourists to see the adaptive reuse process.
- (2) -
- (3) -
- (C) -

Site ownership

- (1) It is necessary to coordinate the site redevelopment with the Bistrica axis (trade spaces, small crafts and services in the town center).
- (2) Single owner and mono-structured production is not appropriate.
- (3) -
- (C) -.

7.4 Planning Recommendations

Considering the unusual workshop settings in Tržič, the planning recommendations could be drafted only during the post-workshop phase. The most significant effort was to translate the outcomes of the discussion (in Slovenian and structured very differently from the previous pilots) into transformation scenarios for the site (start-up, intensive, extensive), as well as to identify the commonly agreed key elements (the key map). For each development path, the key actions related to the structuring issues for the site transformation—ecological network and public open spaces (including potentially relating built-up ones)—are described with text and an attached illustrative plan.

Start-up/temporary activation

In view of a comprehensive transformation of the former industrial area, the reactivation of some of its parts might help, as a priority intervention, to trigger the whole regeneration process. The 'start-up' path envisages specific actions aimed at preparing the area, such as the preservation and gradual reuse of the most prominent industrial heritage buildings on the west end of the area. The building reactivation involves the identification and development of a minimal ecological and mobility/open spaces network, in order to connect the activated parts of the site and the outer areas (Table 7.4; Fig. 7.6).

Table 7.4 Start-up scenario, key issues

Ecological network	*Landscape-environmental recovery of riverbanks*: promote facilities on the banks, pathways, landscaping, etc.
	Tržič Bistrica restoration: manage the flooding risk on the site by creating a buffer zone between the river and the adjacent urban neighborhoods
	Green connections: connect green areas and the river corridor with in-site and off-site built areas, as well as to the hillside landscape
	Blue connections: enhance and better connect the river with the site, especially nearby the industrial heritage buildings
Public open spaces	*Industrial heritage buildings*: preserve and reuse the industrial heritage structures connected to the site 'historical memory' (landmarks) and restore the surrounding open space for public use
	New entrance: open up the west end of the site as a new main entrance (nearby the heritage buildings)
	Public open spaces and pathway network: connect the activated parts of the site and enhance the livability of the area

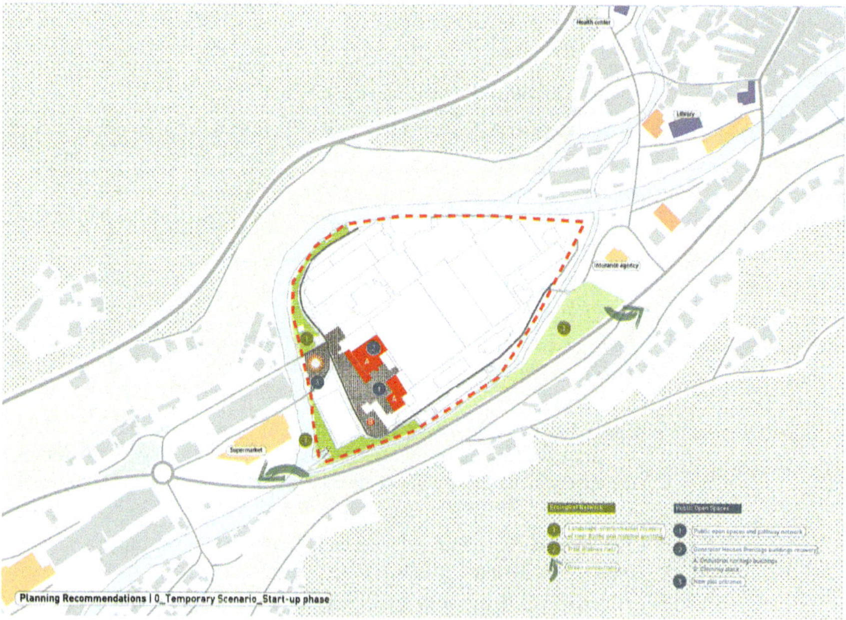

Fig. 7.6 Start-up scenario. © Politecnico di Milano/Interreg project *trAILs*

Intensive development (a)/ ***back to city***

The first 'back-to-city' development path addresses a substantial redevelopment of the site through the reuse and reactivation of most of the existing industrial buildings (either old or recent), suitable to host micro and small local enterprises (handcraft workshops, services, tourism, culture). To this purpose, the site internal layout is slightly refurbished by setting-up of a network of new paths, public spaces and servitudes that integrates coherently the existing structure, while ameliorating the ecological conditions of the whole area (Table 7.5; Fig. 7.7).

Intensive development (b) ***/back to city (hybrid)***

The second 'back-to-city' development path provides a hybrid solution, in which the reuse and reactivation of some of the existing industrial buildings (either old or recent) coexists with a partial removal of other structures and the definition of new ecologically-oriented building zones. The transformation is set according to a unitary objective: a key investor should start a new productive activity on the site, in accordance with natural resources, e.g. by reusing the existing waters system (turbine plant, river and canal, water basin). Just as an example, the transformation could be oriented towards the development of an aquaponic facility (freshwater aquaculture and hydroponic agriculture) (Table 7.6; Fig. 7.8).

Table 7.5 Intensive development scenario, key issues

Ecological network	*Landscape-environmental recovery of riverbanks*: promote facilities on the banks, pathways, landscaping, etc.
	Tržič Bistrica restoration: manage the flooding risk on the site by creating a buffer zone between the river and the adjacent urban neighborhoods
	Increasing of vegetation cover: enhance the semi-natural look of the site by expanding and improving the actual vegetation cover, in line with the built/green ratio of the surroundings
	Green connections: connect green areas and the river corridor with in-site and off-site built areas, as well as to the hillside landscape
Public open spaces	*Industrial heritage buildings*: preserve and reuse the industrial heritage structures connected to the site 'historical memory' (landmarks) and restore the surrounding open space for public use
	Other industrial buildings: adaptive reuse of former industrial buildings by implementing innovative architectural solutions, in order to restore the site attractiveness and enhance its attached culture of technological innovation
	New entrance: open up the west end of the site as a new main entrance (nearby the heritage buildings)
	Tree-lined parking: plant trees in parking lots in order to decrease pollutants and reduce temperature on the ground line
	Public open spaces and pathway network: connect the activated parts of the site and enhance the livability of the area

Fig. 7.7 Intensive development scenario. © Politecnico di Milano/Interreg project *trAILs*

Table 7.6 Extensive development scenario, key issues

Ecological network	*Landscape-environmental recovery of riverbanks*: promote facilities on the banks, pathways, landscaping, etc.
	Tržič Bistrica restoration: manage the flooding risk on the site by creating a buffer zone between the river and the adjacent urban neighborhoods
	Increasing of vegetation cover: enhance the semi-natural look of the site by expanding and improving the actual vegetation cover, in line with the built/green ratio of the surroundings
	Green connections: connect green areas and the river corridor with in-site and off-site built areas, as well as to the hillside landscape
	Blue connections: enhance and better connect the river with the site, especially nearby the industrial heritage buildings
Public open spaces	*Industrial heritage buildings*: preserve and reuse the industrial heritage structures connected to the site 'historical memory' (landmarks) and restore the surrounding open space for public use
	Other industrial buildings: adaptive reuse of a few other industrial buildings by implementing innovative architectural solutions, in order to restore the site attractiveness and enhance its attached culture of technological innovation
	New entrance: renovate the old building at the entrance of the site (new gate of the old factory)
	Public open spaces and pathway network: connect the activated parts of the site and enhance the livability of the area

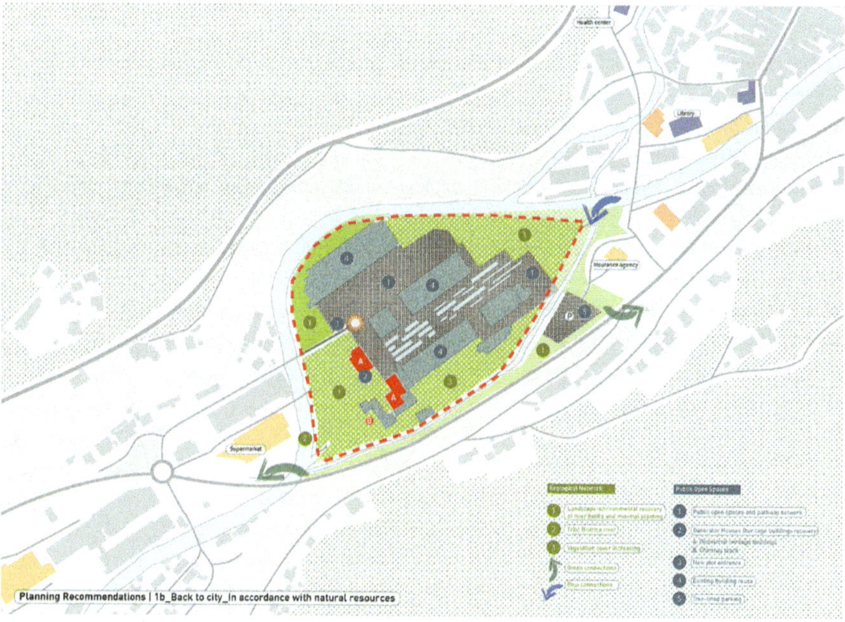

Fig. 7.8 Extensive development scenario. © Politecnico di Milano/Interreg project *trAILs*

Key Map

As shared between the three development paths, and preliminary to their future implementation, the key map integrates and illustrates the commonly agreed issues as identified in the test-design workshop. These include the essential network of public spaces, the environmental system and relevant ecological spaces as well as the most relevant buildings and structures to be preserved and reused. By displaying the basic structure for the future development of the site, the key map is a valuable tool in the hands of the community, a starting point for any possible transformation. The minimum requirements for the site transformation and regeneration, no matter what will be the scenario that the community implement, are listed here below (Table 7.7; Fig. 7.9).

Table 7.7 Key map issues

Landscape-environmental recovery of riverbanks: promote facilities on the banks, pathways, landscaping, etc.
Tržič Bistrica restoration: manage the flooding risk on the site by creating a buffer zone between the river and the adjacent urban neighborhoods
Increasing of vegetation cover: enhance the semi-natural look of the site by expanding and improving the actual vegetation cover, in line with the built/green ratio of the surroundings
Green connections: connect green areas and the river corridor with in-site and off-site built areas, as well as to the hillside landscape
Blue connections: enhance and better connect the river with the site, especially nearby the industrial heritage buildings
Public open spaces and pathway network: connect the activated parts of the site and enhance the livability of the area
Industrial heritage buildings: preserve and reuse the industrial heritage structures connected to the site 'historical memory' (landmarks) and restore the surrounding open space for public use
New building zones: define and delimit unbuilt lots for the site future development, set to host new buildings, activities and functions

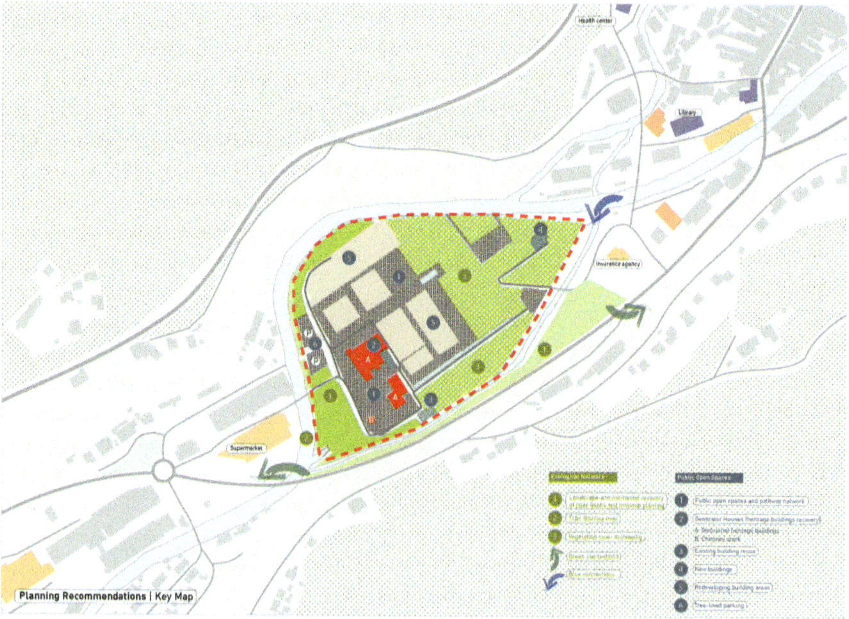

Fig. 7.9 Key map. © Politecnico di Milano/Interreg project *trAILs*

References

Bär K, Kollmann J (2020) EU Interreg project trAILs—Deliverable D.T2.2.6.—Environmental assessment report Tržič. Available at: https://www.alpine-space.org/projects/trails/pilots/_results/wp-t2/d.t2.2.6-environmental-assessment-report-trzic.pdf

Cvenkel H, Brankovič U, Terčon S (2019) EU Interreg project trAILs—Deliverable D.T1.3.2.—Pilot Profile Tržič. Available at: https://www.alpine-space.org/projects/trails/pilots/_results/wp-t1/bpt-profile-official.pdf.

Krošelj M, Pipan T, Marot N et al (2019) EU Interreg project trAILs—Deliverable D.T2.2.2.—Policy assessment report Tržič. Available at: https://www.alpine-space.org/projects/trails/pilots/_results/wp-t2/d.t2.2.2-policy-assessment-report-trzic.pdf

Lanfranchi G, Migliorati L (2020) EU Interreg project trAILs—Deliverable D.T2.2.4.—Socio-demographic assessment report Tržič. Available at: https://www.alpine-space.org/projects/trails/pilots/_results/wp-t2/d.t2.2.4-social-demographic-assessment-report-trzic.pdf

Polin V, Prete V (2020) EU Interreg project trAILs—Deliverable D.T2.2.5.—Economic assessment report Tržič. Available at: https://www.alpine-space.org/projects/trails/pilots/_results/wp-t2/d.t2.2.5-economic-assessment-report-trzic.pdf

Polytechnic University of Milan and Vienna University of Technology (2020) Deliverable D.T2.2.3. Spatial planning and landscape assessment report Tržič. In: Solero E, Vitillo P, Galuzzi P et al (eds) Project trAILs

Chapter 8
Pilots Comparative Review

The chapter provides a comparative evaluation of the activities performed in the four pilot sites, described in the previous chapters, and the achieved results. The evaluation is developed on a site-base and using a set of key indicators derived from the test-design procedure. A comparative table is used to synthesize and visualize the differences and similarities among the pilots. At the end, some critical considerations are expressed with regards to the adaptability of the procedure itself and the influence of contextual conditions.

8.1 Introduction: Co-design Indicators

Following the detailed description of the different activities carried out in each of the four pilot sites, a comparative review becomes necessary to critically evaluate the process as well as to understand the limits and opportunities of the used tools and formats. The evaluation is mainly focusing on the co-design phase—including the setting up, structuring and development of the workshops—as the most relevant step involving the regional stakeholders and the local community. It must be clarified, in advance, that the structuring and organization of the co-design workshop must not be replicated identically in different situations, but it should be carefully calibrated according to the specific social, cultural, economic and legislative context. Only in this way, it will be possible to achieve results whose concrete application and further implementation are indeed realistic, and thus feasible. In addition, the physical setting of the workshop has been excluded from the evaluation. Different options were initially discussed: (a) on-site (within the brownfield site itself), (b) external location within walking distance to the site and (c) 'formal' seminar location within the municipality and/or region. Although the preferred option would have been the first one, better in terms of public visibility and thus attractiveness, the general conditions of the sites (dangerous structures, lack of infrastructures, etc.) and the difficulties in setting-up there a multi-day workshop led to choose the option (b).

M. Modica and E. Solero, *Brownfield Transformation in Fragile Territories*,
PoliMI SpringerBriefs, https://doi.org/10.1007/978-3-031-09083-7_8

The evaluation of the four experiences has been done using a few key indicators, derived from the methodological background on co-design and participatory planning (see Chap. 3), as well as defined based on the four workshops conducted within the project. The indicators are:

- input materials (types, content and characteristics)
- matrix (as comparison/selection tool)
- table setting (thematic or interdisciplinary)
- facilitators (presence and role)
- stakeholders (selection, involvement, sectors, numbers)
- agenda (set-up, duration and sharing)
- key map (aim and use).

For each of the indicators, a very brief description is provided according to the actual experience. This allows to quickly and easily compare the various aspects in the focus, while also providing a focused and compact overview on the activities carried out on each pilot site. The four experiences are finally compared, using the indicators and a summarizing table, to draw a critical conclusion on the effectiveness and adaptability of the co-design workshop.

8.2 Eisenerz (Austria)

Pilot area

Former Münichtal blast furnace plant.

Input materials

To facilitate the discussion among the participants, allowing also non-experts to understand a design-based language, a set of helpful materials was made available at the workshop tables:

- *site assessment reports*: a mixed text-graphic printed document obtained by just assembling the five thematic assessment reports compiled by the research partners (policy analysis, spatial planning and landscape assessment, social assessment, economic assessment, environmental assessment);
- *blank map*: topographic (zenith and axonometric) base maps focusing on the pilot area and its surroundings, serving as support for sketching the first strategic ideas for the site transformation;
- *illustrative poster of transformation scenarios*: the suggested scenarios—renaturation, greenhouse, recycling—were used as main reference for the discussion with the stakeholders, as well as the basis for the development of new proposals. Topics could be added through sketches and by attaching post-its, easily combined or removed. The scenarios were not meant as project proposals, but they just had to stimulate the confrontation;

- *socio-economic booklet*: a short infographic printed brochure providing the main socio-economic data and findings;
- *environmental questionnaire*: a series of images relating to natural elements present on-site and questions to which each participant had to answer, relying on the sensations/perceptions aroused by the images.

Matrix

No matrix was designed nor used, because the tool was not yet defined in its main features and uses. In addition, TUW already prepared three possible transformation scenarios for each of the workshop tables.

Table setting

As first suggestion, it was proposed to work with three thematic tables and the corresponding stakeholders and regional experts:

Table 1: society, culture and economy.

Table 2: ecology and landscape.

Table 3: spatial planning and urban development.

Considering the number and type of registered stakeholders, as well as the three proposed scenarios by TUW, it was decided to shift to three interdisciplinary tables each focusing on one specific scenario, distributing the stakeholders accordingly together with research partners, one moderator and one reporter.

Facilitators

No facilitators were involved. Team members from TUW acted as facilitators at the tables, including simultaneous German-English translation and reporting.

Stakeholders

The involvement of stakeholders represented a real challenge, mostly due to the difficulties in conveying the whole region on working on one single site. Around 50 stakeholders and observers were contacted personally, aiming to involve the relevant target groups. In the end, only 17 stakeholders attended the workshop.

A list of questions was prepared beforehand, to spark the discussion among the stakeholders on the relevant topics.

Agenda

The workshop lasted a total of three days. The first day included a second, short site-visit (together with a few stakeholders), to which followed a round discussion about the content of the dossiers and preparatory activities in the view of the working session with the stakeholders and observers (second day). The third day was dedicated to the workshop synthesis, next steps and steering group meeting. The workshop agenda was set up jointly by TUW and POLIMI and shared with all partners. It was clear and well detailed. It had been revised in the evening of the first day as regards the table setting.

Key Map

No key map was produced during the workshop, as the final discussion did not allow to merge the three scenarios into a shared single one. The key map was instead produced in the post-workshop phase, based on the critical review of the workshop outcomes by the project partners.

8.3 Borgo San Dalmazzo (Italy)

Pilot area

Former Italcementi cement plant and related quarries.

Input materials

After the first experience in Eisenerz, it was decided to improve the quality and focus of some of the input materials, highlighting the most relevant issues to ensure an effective discussion. The choice to use, at the working tables, some tools rather than others, or to introduce new tools, was motivated by the need to simplify the discussion process as much as possible, in order to provide all the participants with clear data and understandable arguments for concrete and relevant issues (e.g. estimation of project costs, procedures and feasibility with regards to existing constraints, temporal development, etc.). The input materials were:

- *site assessment dossier*: a mixed text-graphic document resulting from a reasoned summary of the thematic assessment reports compiled by the research partners (policy analysis, spatial planning and landscape assessment, social assessment, economic assessment, environmental assessment);
- *cards*: thematic cards produced by the research partners, each describing through one image and one sentence a relevant issue emerged from the thematic assessments and concerning the transformation potential of the site. The cards are used to facilitate the selection of key topic among the stakeholders, for the discussion;
- *blank map*: topographic (zenith and axonometric) base maps focusing on the pilot area and its surroundings, serving as support for sketching the first strategic ideas for the site transformation;
- *design proposals*: as design input, TUM and POLIMI independently produced some design proposals for the site transformation, together with students in landscape architecture (TUM) and architecture and urban planning (POLIMI). The proposals were presented by means of two separate booklets, available at the workshop.

Matrix

Considering the amount of design proposals (5 by TUM and 4 by POLIMI), a matrix was created to compare the projects and highlight the most significant emerging issues for the discussion with stakeholders and observers. At the working tables, the matrix mainly served to provide at the same time an overview on the different design alternatives as well as a quick and detailed reading of components and issues in each single proposal.

Table setting

Three thematic tables, each focusing on all the projects contemporarily, were set up, associated to a color and an expert moderator:

Table 1 (green table): environmental and landscape issues.

Table 1 (red table): socio-economic issues.

Table 3 (blue table): territorial and planning issues.

To ensure a transparent and understandable communication among the participants, the discussion at each table was constantly reported (represented) on a poster by means of sketches and notes. In this way, most of the involved participants was able to integrate their own point of view, while also keeping an overview on the table discussion results. Furthermore, this procedure ensured that no forms of supremacy (e.g. who speaks better or has a relevant role) could take place.

Facilitators

External and expert facilitators were involved. Thc facilitator team was composed by one general moderator, a visual designer and, at each table, a group moderator and a translator (Italian-English).

Stakeholders

The selection of stakeholders to be involved was motivated by the need to have a wide variety of local and regional interest groups directly or indirectly affected by the future site transformation. In particular, besides institutional stakeholders and experts, it was decided to invite representatives from culture, health, industry, education and social innovation sectors. On 51 invited stakeholders, 19 joined the workshop for one full day. Among those, all relevant sectors of the local community were represented.

Agenda

The workshop lasted a total of three days. The first day was devoted to preparatory activities, as well as to the revicw of the assessment dossiers and the definition of expected outcome for the following day. The second day consisted in the discussion with stakeholders, observers and experts on the site's future. The third day was dedicated to the workshop synthesis, next steps and steering group meeting. The workshop agenda was set up by POLIMI and shared with all partners. It was clear and well detailed.

Key Map

During the second day (discussion with stakeholders), a visual designer collected the emerging issues from the tables, rendering those into an artistic/pictorial reproduction, which was then left to the community as a visionary artwork. No key map was produced at the end of the workshop, due to difficulties in reaching a common view among the three tables on the site's future priorities. Three different development paths were then outlined in the post-project phase, following the preliminary results, and from those a key map as start-up phase.

8.4 L'Argentière-La-Bessée (France)

Pilot area

Former Pechiney aluminum smelter.

Input materials

Based on the previous experiences, the input materials were further simplified to allow a straightforward and clear discussion. The materials used were:

- *site assessment poster*: printed A0 panel summarizing and harmonizing the most relevant issues emerged from the thematic assessments compiled by the research partners (policy analysis, spatial planning and landscape assessment, social assessment, economic assessment, environmental assessment);
- *map of public and private properties*: provided by CAUE 84, the map was a very useful add-on to the workshop, considering the high fragmentation of the land properties within and around the site;
- *aerial zenithal photo*: aerial photograph of the site, including the Sablonnières activity zone and AFP brownfield;
- *design proposals*: as design input, TUM produced three design proposals for the site transformation, together with students in landscape architecture (TUM). The proposals, on A0 poster format, presented different degrees of a rather radical transformation for the site;
- *framework questions*: a list of questions, on paper, was prepared beforehand to enable the exchange among the workshop participants. The questions focused on the level of knowledge of the brownfield background and current situation, as well as on the perception of it in relation to the town and the Durance valley. The questions were directly used at the working tables.

Matrix

The matrix was not produced nor used, as the input designs were already in the number of three and no further comparison/selection was necessary.

Tables setting

Three interdisciplinary tables (stakeholders, experts, project partners) focusing each on one single design proposal. No exchange of participants or projects during the discussion time.

Facilitators

Due to the expertise in planning procedures and community engagement, CAUE 84 together with CAUE 05 acted as facilitators. Each working group was provided with one moderator and one translator (French–English).

Stakeholders

The selection and distribution of local and regional stakeholders around the working groups was set-up with the aim to mix profiles and expertise as much as possible, in order to encourage an exchange between often distant viewpoints and approaches to brownfield transformation. The allocation of the involved stakeholders—21 in total—to the tables was prepared in advance. At the tables, priority was given to the stakeholders for sharing their opinion and suggestions, while the project partners and other experts only provided complementary insights (not influencing).

Agenda

The workshop lasted a total of three days. The first day was devoted to preparatory activities, as well as to the review of the assessment dossiers and the definition of expected outcome for the following day. The second day consisted in the discussion with stakeholders, observers and experts on the site's future. The third day was dedicated to the workshop synthesis, next steps and steering group meeting. The workshop agenda was set up by CAUE 84 and CAUE 05, with the support of POLIMI and TUM, and shared with all partners. It was clear and well detailed.

Key Map

A draft key map (in form of a 'programmatic plan') was outlined already at the end of the second day, after a collective review of the preliminary results from the working tables in an open plenary session with all the participants. The draft was then finalized in the post-workshop phase.

8.5 Tržič (Slovenia)

Pilot area

Former BPT cotton mill.

Due to the Covid-19 pandemic, the planned on-site activities in Slovenia had to be carried out in online or hybrid mode. In particular, the test-design workshop was conducted in hybrid mode, with only the local project partners and a few stakeholders meeting in presence. The workshop thus encompassed both remote and face-to-face sessions, the firsts meant to enable a minimum exchange between the project partners and the stakeholders (English and Slovenian languages) and the seconds to let the local group to deepen certain issues and questions (Slovenian only). The stakeholders were grouped in one single room with a projector and microphone, while all the project partners (except the Slovenian ones) were connected via GoToMeeting.

Input materials

- *site assessment dossier*: compact printed booklet summarizing the potentials and problems emerged from the thematic assessments compiled by the research partners (policy analysis, spatial planning and landscape assessment, social assessment, economic assessment, environmental assessment);
- *design proposals*: as design input, TUM produced some design proposals for the site transformation, together with students in landscape architecture (TUM).

Matrix

Considering the amount of design proposals (6 by TUM), a matrix was created to compare the projects, highlight the most significant issues for the discussion with stakeholders and observers and, most important, to select 3 design proposals for the workshop. The matrix thus served to provide at the same time an overview on the different design alternatives as well as a quick and detailed reading of components and issues in each single proposal.

Tables setting

Due to the circumstances, only one working table could be set up physically, joined by the stakeholders and the local research partners (UL and BSC KRANJ).

Facilitators

UL, with the support of BSC KRANJ, moderated the discussion at the table as well as the remote communication with the rest of the group (translation Slovene-English included).

Stakeholders

Ten local and regional stakeholders attended the workshop. Key questions were prepared in advance by UL and the other research partners, in order to stimulate the discussion with the stakeholders on specific topics.

Agenda

The workshop lasted a total of two days. The first day (hybrid mode) was devoted to a brief presentation of the assessments and a roundtable discussion with stakeholders, observers and experts on the site's future. The second day (online only) was dedicated to the workshop synthesis and steering group meeting.

Key Map

No key map was outlined during the workshop. The outcome consisted in a sort of updated matrix which compared the scenarios with the problems/opportunities and insights emerged during the discussion.

8.6 Critical Considerations

The synthetic review of the test-design procedure carried out in the four pilot sites shows many common characteristics among all the experiences, but also several key differences. In all the cases, the procedure was indeed based on the active involvement of the local community and the regional stakeholders. Guided by the project team and the planning experts, the communities were called to critically evaluate a set of transformation proposals (test-designs) and to identify, through both group and collective discussion, the most relevant issues for the site's future development (translated then into planning recommendations). The design approach assumed various formats (scenarios, masterplans, etc.), depending on previous experiences and adapting to the specific local conditions, including the number of involved people, the sectoral variety within the stakeholders' group and the complexity of the transformation site itself. Concerning in particular the co-design workshops, the original methodological framework (see Chaps. 2 and 3) was therefore constantly modified pilot after pilot, never resembling the same exact format in more than one site.

In Eisenerz, the workshop set-up and organization followed the principles of "test-planning", an informal planning procedure specifically designed for planning experts. Basically, test-planning aims to produce legally binding agreements for the site's future transformation through the interdisciplinary review of different project proposals, as in a sort of simulated competition. However, considering the limited number of stakeholders and their low acquaintance with complex planning issues and languages, the design proposals were indeed rather simple transformation scenarios, presenting three alternative futures for the Münichtal site both in terms of program and project. The *pros* of this approach consisted in the strong legitimation of the output (concrete agreement among decision-makers), which however required more space for discussion than what was really planned (time constraints). The *cons* were mainly related to the real difficulties in adapting and implementing this process to marginal and fragile territories in the broad sense, where the required highly specialized planning experts are indeed lacking. Eisenerz was the first among the regional workshops. In this regard, it was a real experiment (a 'pilot' within the pilot itself) for what came next. A crucial issue was the successful involvement of regional stakeholders and the local community, which turned out to be a very challenging task for the regional partners. It was learned, soon, that a sound promotion activity needs to be done regionally and locally, to ensure at least a full day participation at the workshop without 'promising' too much personal profits.

In Borgo San Dalmazzo, the workshop procedure resembled much more what is well-known as participatory planning, also an informal planning process but open to a variety of local stakeholders (not only planning experts). This approach aims to generate a shared vision for the site's future through an open listening process, in which the local communities and representatives are encouraged to share their views, needs and knowledge with reference to the problem. Differently from the previous experience, the working groups were here strongly thematic-oriented. Motivated by the type of involved stakeholders (not many experts but more representatives of the civil society), the thematic orientation of the workshop tables allowed to deepen certain questions and issues apparently not strictly related to the brownfield site, but indeed influencing/influenced by its future transformation. The *pros* of this procedure consisted mainly in the great inclusiveness of local interests, which translated in a strong community-legitimation of the outputs. At the same time, the 'non-binding' character of the outputs (a vision, not a plan) constituted a relevant *cons*. The approach followed in Borgo San Dalmazzo gave more relevance to the process itself, rather than to the design projects, with the aim to identify a listening procedure valuable for many similar Alpine territories and contexts affected by brownfield transformation.

Based on the previous two experiences, in L'Argentière-la-Bessée it was developed and tested a hybrid approach, in which the outcome of the workshop was represented by a mutually agreed list of key planning issues (minimum requirements) for the future development of the site. The special conditions of the French experience—namely the highly mixed group of stakeholders (local interests, planning experts and regional representatives) and the planning-oriented moderation by the experienced regional partners CAUE 84 and CAUE 05—allowed to integrate the most successful elements from both test-planning and participatory planning, resulting in a real planning scheme though understandable even by non-experts, and inclusive of many non-planning issues.

In Tržič, the procedure was strongly affected by the pandemic restrictions. A hybrid approach was used also here, although the discussion was mostly limited to the local group (in presence) of stakeholders and Slovenian partners. Nevertheless, even under these difficult circumstances the workshop proved to be rather successful, at least at the local level.

To allow a clear and detailed overview of the differences and similarities among the four experiences, the indicators have been grouped and synthesized in Table 8.1. The latter highlights when thematic (T) or interdisciplinary (I) tables were used, if the comparative matrix was used as a key tool, if facilitators were present, how lasted the workshop and if ended with a key map (synthesis) or not.

Once more, it is important to underline that there is no perfect recipe for a perfect workshop but, as the table shows, it is necessary to adapt the process and procedures to the context, the specific situation and the actors involved.

Table 8.1 Comparative overview of the four test-design experiences

Pilot site	Tables		Matrix	Facilitators	SH	Ws duration		Agenda	Key map	
	T	I				1d	< 1d		During Ws	After Ws
Eisenerz		x			17		x	x		x
Borgo SD	x		x	x	19	x		x		x
L'Argentière-la-Bessée		x		x	21	x		x	x	
Tržič*		x	x		10	x		x		x

* hybrid, T = thematic, I = interdisciplinary, SH = stakeholders (participating), Ws = workshop session with stakeholders, d = full day

The organization and set-up of the working tables can be thematic (Borgo San Dalmazzo) or interdisciplinary (other pilots). The thematic approach allows a deeper exchange between competent and interested actors, as well as a strong link between the site transformation and specific, sectoral matters. The discussion can be extremely fruitful and concrete, but for a successful outcome it is necessary to involve a high number of stakeholders from many different sectors/levels. The interdisciplinary approach, on the other hand, allows a direct exchange between different expertise, which is often lacking in planning procedures happening in marginal territories. However, this modality requires a lot of preparation in terms of moderation and communication, to be able to achieve concrete results in a limited amount of time.

The comparative matrix, used in Italy and Slovenia, revealed to be a very useful tool for a twofold reason. It allowed to understand and compare the different design proposals by reading and simplifying those through their main thematic layers (transport and accessibility, open spaces, built spaces, ecologically sensitive areas, etc.), while also easing the selection of those design proposals most useful to the workshop purpose.

The participation of skilled facilitators in the Italian and French experiences was helpful to guide the discussion and improve the communication on the tables and between them. Facilitation has proved to be essential in such workshops, as it allows a fair and inclusive involvement of all the stakeholders in the discussion, while also monitoring and reporting the debate as an 'external' viewer. At the same time, as demonstrated in both experiences and especially in France, the facilitators should be well acquainted with the discussion topics and issues on the table. As learned, it is crucial to involve the right stakeholders (type and number) in order to have a fruitful discussion and concrete, useful/usable outcomes.

Concerning the duration and the agenda, the four experiences shared mostly the same timing and structuring. In Eisenerz, the phase with stakeholders lasted more than one day, while in Italy, France and Slovenia they were invited for one day only. In Italy and France, in particular, the stakeholders remained within the workshop for one full day, reaching the final synthesis phase together with the project partners and experts. This was extremely helpful to draft a good conclusion. Considering the difficulties in having the stakeholders for more than one day, it is essential to set up compact working sessions with clear instructions on how to proceed and what to achieve (a clear and shared agenda).

The key map, finally, as the main outcome of the workshop, was variably interpreted and produced in each experience. Basically, the key map served to include and represent the shared agreements among the workshop participants on the sites' future transformation. Drafting these conclusions, and the key map too, within the workshop itself, together with the stakeholders and communities, is the best solution. This was the case of France, where the great and focused participation enabled so, while in Austria, Italy and Slovenia the key map had to be outlined in the post-workshop phase. In all the four cases, the key map was then shared with the workshop participants (experts, communities and stakeholders) to get feedback on the actual understanding and accuracy in respect to the workshop discussion. Only in this way the key map can be used as the basis for concrete planning recommendations useful to orient the future of the site.

Chapter 9
Conclusions and Recommendations

This conclusive chapter reviews the scope and aim of the book in light of its content. It discusses, in particular, the test-design procedure as the key output of the project as well as the project general approach, as a combination of action research and transnational cooperation. A first section focuses on the test-design characteristics, advantages and preconditions for a successful implementation. A second section, finally, highlights the limits experienced in the project as well as the main learned lessons, providing recommendations for future, similar experiences.

The trAILs project has first explored, jointly and interdisciplinary, the emerging challenge of brownfield transformation in mountain contexts shaped by multi-level fragilities. To do so, the project developed a specific planning methodology defined as 'test-design', which combines design-driven research with participatory approaches on different brownfield sites across respective eco-cultural regions within the transnational Alpine space. Following the key principles of action research, trAILs actively involved many local communities, institutions and stakeholders in the pioneering exploration of Alpine brownfield transformation, fostering a productive and enriching dialogue between experts and citizens, as well as transnational and local actors and networks. The project embraced deliberately a procedural approach 'from-project-to-process', recognizing the plural condition of marginal areas, the existing capabilities and resources, and especially the uncertainty of future developments for what concerns large-scale spatial and territorial transformations such as brownfield revitalization. This multi-faceted and challenging condition, differently 'interpreted' at the regional scale through the pilot sites, called the project consortium to put in place a shift in both the thinking and the operative/methodological pattern; a necessary precondition to explore and develop new approaches suitable to the challenge on stake.

The test-design procedure, which can be considered at the same time the most significant output of the trAILs project and *the* project itself, has succeeded in establishing a transnational framework for mutual understanding and learning, a valuable occasion of dialogue—formal and informal—between different yet interlinked know

M. Modica and E. Solero, *Brownfield Transformation in Fragile Territories*,
PoliMI SpringerBriefs, https://doi.org/10.1007/978-3-031-09083-7_9

hows and interests. The key advantages of test-design in relation to complex spatial and territorial transformations in weak contexts can be summarized as following:

- to allow communities, stakeholders and experts to meet face-to-face around a working table, thus creating the best conditions for direct and interdisciplinary communication, as well as for an exchange of opinions with full transparency, understanding, and mutual trust;
- to co-identify and agree on the most critical key issues for future spatial transformations, such as increasing social infrastructures, fostering a more sustainable economic development, improving ecological and environmental systems, reducing risk, protecting and managing architectural heritage and historical values;
- to provide concrete and feasible planning recommendations for the future implementation, a valuable instrument in the hands of the local communities involved.

Through the project focus of brownfield transformation and redevelopment, the test-design procedure has proved its absolute relevance in complying to the need of usable (not abstract) knowledge expressed by local communities. The project outlined in fact appropriate and sustainable strategies for the regeneration of marginal areas at risk, thus contributing substantially to implement territorial and community resilience. However, the project experience suggests that to be really effective to its scope, test-design must satisfy a few key conditions or requisites.

At first, test-design should be set as a true transdisciplinary exchange, not just integrating different disciplinary views and expertise (interdisciplinarity) but more as a joint convergence towards a shared objective, thus setting a broader conceptual and operative framework for intervention. Test-design brings into the same process, or flow, many participants from a wide knowledge spectrum, necessary to address brownfield transformation in weak contexts, such as spatial and regional planners, architects and landscape architects, sociologists, public policy analysts, ecologists and environmental scientists, economist, each with its own vocabulary and toolbox. It is essential to allow an open and interactive confrontation among these experts, fostering 'cross-fertilization' through co-study and co-evaluation of the same questions. Certainly this is not an easy operation, but the test-design format—from the site visit to the final planning recommendations—is designed to help reaching transdisciplinary frontiers in both the intentions and the results.

Secondly, test-design should carefully and openly listen to the local communities affected by spatial challenges and changes. The interaction with local inhabitants, institutions and represented interests is fundamental to ensure the concreteness as well as the acceptability of identified solutions. This listening process is constantly implemented in the test-design procedure through face-to-face meetings, individual conversations and video-interviews, thus enabling the community to feel as part of the solution. The active participation of local stakeholders must be carefully planned since the beginning of the procedure and increasingly nourished as the confrontation goes on. Only in this way a participatory approach such as the one expressed by test-design can enhance the planning capabilities and the territorial responsibility of local communities.

Thirdly, test-design must provide generative knowledge, in the sense that the achieved solutions and results should foster long-term strategies and activate new processes beyond their own scope. Test-design does not provide detailed plans or readymade designs for complex spatial transformation, but it helps instead to bring together communities and experts, local and supra-local knowhow, in the co-identification of transformation goals and co-design of future-oriented strategies to reach those. Design itself works as a valuable communication tool to support this co-development process, being able to highlight the emerging questions and to stimulate reality-based discussions and live confrontation. The main output of the test-design, i.e. the planning recommendations, summarize through a design language—clear and accessible to 'non-experts' too—the future perspectives for the site transformation derived from co-identified key issues. As a type of generative knowledge, the recommendations should be able, by themselves only, to activate long-term transformation processes in marginal or weakened communities and territories. Furthermore, the same knowledge should incorporate elements of replicability and adaptability to allow an effective transferability between near and far situations with same problems and characteristics.

9.1 Limits and Lessons Learned

In the framework of the trAILs project not only a new methodological approach to complex planning issues such as brownfield transformation was developed and tested, namely the test-design procedure, but broadly the integration between action research and transnational cooperation was experimented and successfully realized. The scope of this closing paragraph is to highlight the limits evidenced in the project as well as the lessons learned from it.

Concerning the former, these are clearly related to the wide differences (geographical, cultural and professional) among participants and territories. The limits can thus be summarized as:

- the difficulties experienced in selecting the most representative stakeholders for the site/challenge on stake, as well as in actively involving them in the listening, assessment and design evaluation process over several months—as required in the test-design scheme;
- the substantial administrative, technical and legislative differences among the involved regions and/or sites, which influence the test-design outcomes and the content of the planning recommendations. Although belonging to the same macro-area (the Alps, in this case), the local discrepancies in terms of environmental, economic and social conditions are hardly comparable or assimilable at some kind of model;

- the difficulties in activating and promoting brownfield transformation and regeneration processes in weak economic contexts, characterized by poor real estate market, limited pressure for reuse and lack of key investors. The territorial and economic 'marginality' of the mountain territories in the focus of the project pushed for exploring even radical future scenarios such as the complete renaturalization of the sites;
- the obvious multi-level difficulties in the communication among the participants. The fact that national languages are mostly spoken locally, by communities and stakeholders, implies that good interpretation skills are necessary to enable the exchange. Technical languages are often very distant too, between disciplines as well as international/global-oriented actors (experts) and proximity-oriented ones (local stakeholders).

Beyond these evident limits, which can indeed be managed through a careful preliminary planning of the procedure, the trAILs project brings forward some inspiring lessons which were learned by the participants 'in the making' (learning by doing).

The first learned lesson relates to the importance of a site-oriented approach. Any brownfield site, as well as the afferent territory and community, possesses unique qualities, characteristic and problems as well. While it is true that sites sharing the same historical and productive background bring similar questions (e.g. certain industries causing specific contamination problems), it is equally valid that each site is shaped by locally specific challenges, which derive from a mix of site-related and contextual conditions. A site-oriented approach, as the one implemented in trAILs and fostered by the test-design methodology, is therefore essential to be able to produce valuable, usable and effective results. Furthermore, focusing on the site specificity helps involved communities and stakeholders to be aware of their own situation, and thus to contribute to its solution.

The attitude of local communities is another relevant aspect. Working with fragile contexts such as mountain regions and open questions such as brownfield transformation require to set up a planning procedure capable of engaging local interests and knowhow. The approach tested in trAILs was strongly based on the active involvement of communities in co-identifying future strategies for their own sites. This aspect worked out quite differently from site to site, depending on the level of information, engagement and 'urgence' felt by the community. The contribution of local representatives and stakeholders in the process helped, when substantial, to outline feasible and realistic development scenarios for the sites, building a positive attitude towards transformation and a resilient imagination of the future.

Lastly, the usefulness of practice-oriented, action research in tackling emerging territorial questions. Through the test-design framework, the trAILs project demonstrated how relevant is the successful integration of experts and non-experts in contemporary spatial and territorial planning. Test-design offers indeed a valuable platform for sharing knowledge, increase mutual understanding and from there generate co-designed solutions. Since the beginning the project aimed at producing usable and reality-based outputs, to contribute to both local challenges (the four pilot sites) and transnational strategies (the Alpine-wide transferability).

As a closing statement, we as authors would like to encourage scientists and professionals to make their own expertise at service of common spatial challenges, such as the one addressed by trAILs, being ready to dialogue with territories and settled communities and learn from them, every time.

The manufacturer's authorised representative in the EU is Springer Nature Customer Service Centre GmbH, Europaplatz 3, 69115 Heidelberg, Germany. If you have any concerns regarding our products, please contact ProductSafety@springernature.com

Printed and bound by CPI Group (UK) Ltd, Croydon, CR0 4YY
28/09/2026
02224079-0006